IT WAS ROCKET SCIENCE®

Bill – You are too young to retire!

Tom Taormina

Tom 5/20/14

Published by Motivational Press, Inc.
7777 N Wickham Rd, # 12-247
Melbourne, FL 32940
www.MotivationalPress.com

Copyright 2014 © Tom Taormina

All Rights Reserved
No part of this book may be reproduced or transmitted in any form by any means: graphic, electronic, or mechanical, including photocopying, recording, taping or by any information storage or retrieval system without permission, in writing, from the authors, except for the inclusion of brief quotations in a review, article, book, or academic paper. The authors and publisher of this book and the associated materials have used their best efforts in preparing this material. The authors and publisher make no representations or warranties with respect to accuracy, applicability, fitness or completeness of the contents of this material. They disclaim any warranties expressed or implied, merchantability, or fitness for any particular purpose. The authors and publisher shall in no event be held liable for any loss or other damages, including but not limited to special, incidental, consequential, or other damages. If you have any questions or concerns, the advice of a competent professional should be sought.
Manufactured in the United States of America.

ISBN: 978-1-62865-083-9

CONTENTS

About the Author 5
Foreword 7
Volume I We (Should) Learn From History 13
Chapter 1 Houston, We Have a Problem! 15
    Quality Assurance 16
    We Did What We Were Trained to Do 16
    Splashdown and Lessons Learned 18
Chapter 2 My Story 21
    JFK 22
    The Foundation of The Apollo Business Model 24
    Farewell to Corporate America 26
    The Entrepreneur 27
    The Glory Days of Dell Computer 29
    The FS 9000 Folly 30
    Proving My Theorems 31
Chapter 3 The Rise and Fall of the American Business Empire 36
    You Must Make a Friend of Horror 36
    The Dryer 37
    The Heart Attack 39
    The Big White SUV 42
    Driving School 45
    Identifying the Trend 47
Chapter 4 Headlines of Shame 50
    Family Perishes in House Fire 51
    Another Family Perishes in House Fire 51
    Man Dies In House Fire 52
    Yet Another Family Perishes in House Fire 52
    Youngster Electrocuted While Using a Battery Charger 53
    Even Yet Another Family Perishes in House Fire 54
    Oil-Field Worker Dies in Well-Head Accident 54
    Child Dies from Administration of Incorrect Drug 55
    Aircraft Component Manufacturer Recalls Jet Engine Parts 55
    Woman Seriously Injured In Bicycle Accident 56
    Woman Sustains Injuries after Being Ejected From a Moving Golf Cart 57
Chapter 5 History of the Twentieth Century 59
    The Transition from Achievement to Apathy 71
    Necessity IS the Mother 72
    The First Theorem of Human Rocket Science 73
    The Second Theorem of Human Rocket Science 73
    Back to the Evolving Century 75
    The Second Half of the Century 77
    The Vietnam Connection 78
    The Yuppies 79
    Enter the Personal Computer 80
    The Third Theorem of Human Rocket Science 82
    Becoming Connected 85

Volume 2 Perfecting Mediocrity 87
Chapter 6 Public Schools: The Foundation of Mediocrity 88
Chapter 7 The Media: Our mirrors or our teachers? 96
    Song Lyrics 97
    Television 117
    Commercials 121
Chapter 8 The Terrible Teens 131
    Proving the Theorem 135
    Spartacus 136
    Homeland 138
    Californication 139
    Dexter 140
    House of Lies 141
    Newsroom 141
    Shameless 142
    Does Life Mimic Art or Art Mimic Life? 144
Chapter 9 Living La Vita Vicarious 148
Chapter 10 Successfully Avoiding Personal Accountability 158
    Definition – Baby Boomers 160
    Definition – Generation X (The 13th Generation) 161
    The Boomer Workers 170
    The Gen-X Workers 171
    The Baby-Boomer Professionals 172
    Gen-X Professionals 175
Volume III The Pathological Business Model: Management versus Leadership 178
Volume IV The Kairos Moment: Do we Really Learn from History? 186
Volume V The Human Lessons from Project Apollo 193
Chapter 11 Applying the Lessons Learned 195
Chapter 12 DO NO HARM 207
    The FMEA and FBP Approach to Do No Harm 210
Volume VI Prelude to The Apollo Business Model (ABM) 212
Chapter 13 Root Cause Analysis 216
Chapter 14 Seven Alpha 220
Volume VII Implementing The Apollo Business Model 225
    Vision 226
    Mission 230
    Values 232
    Leadership 234
    Process 235
    Boundaries 239
    Metrics 241
    Consistency 243
    Achievement 245
Chapter 15 The Apollo Business Model Self-Assessment 247
Chapter 16 One Cooler Tool 250
    Epilogue 252

ABOUT THE AUTHOR

Tom Taormina was one of the first Quality Control Engineers when the discipline began its evolution in the 1960's. He was certified in the field by The Ford Motor Company and by NASA at the Johnson Space Center in Houston, where he worked in the Mission Control Center for 14 years. His pioneering work on Project Apollo started a 40-year career, working with more than 600 companies, to amass an extraordinary knowledge of quality systems and how leadership and human accountability contribute to business success or failure.

Tom is the author of ten books on quality, process control, and human dynamics.

Five of his books were published by Prentice Hall and four of his most recent books are seminal works on advanced quality management techniques, developed by his group for Dell Computer. His last book, Foreseeable Risk, is a compilation of a decade of work as an expert witness in products liability and organizational negligence into a desk reference for business leaders to use to avoid litigation.

His writings have been published on five continents and have been translated into Chinese, Spanish, Portuguese, and Korean.

The author witnessing a test in Mission Control circa 1971

Put it before them briefly so they will read it, clearly so they will appreciate it, picturesquely so they will remember it and, above all, accurately so they will be guided by its light.

--Joseph Pulitzer

This book is dedicated to Grady Ferguson, my most trusted friend, business partner and the only person who tirelessly encourages and helps me to continually raise my business and personal standards of excellence. Together we are known as "the two rocket scientists."

FOREWORD
BY ASTRONAUT JACK LOUSMA

Among America's major technological and manufacturing triumphs of the Twentieth Century, one must include the mobilization of World War II and the NASA Apollo program landing on the Moon. Both Federal programs were inspired by international threats from abroad; one military and the other civilian. Moreover, both represented challenges to which America was forced to respond to maintain world leadership in the maintenance of either personal freedom or technological superiority. Both also required similar leadership and management qualities by individuals and organizations across the nation. Their responses set the standards and derived the techniques for which future generations could organize successful projects in both the public and private spheres of innovation, development, testing, and performance.

With respect to the Apollo program, the leadership began in 1961 with President John F. Kennedy's challenge to land Americans on the Moon and safely return them to Earth by the end of the decade. NASA was charged with the responsibility to accomplish this goal and identified a handful of established leaders (not a plethora of bureaucratic committees) to plan, structure, and execute the President's mission on schedule and within budget. Thus, the Mercury, Gemini, and Apollo programs were sequentially arranged to learn what it took to physically and operationally launch into space, survive in that environment, and return to Earth safely. Furthermore, we had to learn how to design rockets, spacecraft, and space suits, as well as, learn how to perform outside the spacecraft, conduct rendezvous and docking, make precise landings, lengthen the time we could survive, and do useful work in zero-gravity and on the Moon.

To accomplish the many tasks within the overall plan, the art of "Systems Engineering" was devised and the principles and techniques of "Project Management" were defined. The principles and techniques of Project Management were used at every level of the Apollo Program to produce space hardware, ensure safety, develop operational procedures, and integrate all elements into a single plan to finish on schedule and within budget. In fact, the first Moon landing in July, 1969 was well within the deadline despite an approximate 18-month delay following the tragic Apollo 1 fire in 1967. Furthermore, the Moon flights were so successful that the last three missions, Apollo 18, 19, and 20, were cancelled because program goals and objectives were achieved sooner than expected.

The key point to be made, as well as one of the objectives of this book, is to demonstrate that the principles of Project Management applied in the Apollo Program are not unique, that is, these same techniques can, and should, be applied to any other public or private sector program to produce results on schedule and within budget. Today, unfortunately, we are experiencing, especially some elements at the Federal level, the wasting of large expenditures of money, time, and effort on programs of national significance by people with no knowledge of the existence of principles of Project Management, much less their responsible applications to achieve positive results.

Despite its importance, Project Management is only part of a successful development program. Some of the most important elements, and there are many more, include leadership, teamwork, risk/reward, personal qualities, commitment to excellence, and a positive attitude. The Project Manager usually works for the leader and manages "things", like schedule and budget, while the leader leads people. A common pitfall is that management is often emphasized over leadership.

In Project Apollo, the leaders were experienced pioneers and entrepreneurs who developed a vision, articulated it to the team, gave clear direction, helped the players run, and kept them focused on the vision. They avoided micromanaging the team and only intervened when the desired outcome was not being served. The leader hires the best team he can find even though some of the players may be more expert in various fields than the leader.

The Apollo team included both government and private-sector organizations with their engineers, scientists, flight controllers, astronauts, tradesmen, technicians, and a host of other disciplines all working together to achieve a common goal. A winning team needs to work in an environment in which all can contribute and also succeed individually as the team succeeds collectively. Everyone shares in the success. Competition between players, within the bounds of cooperation and collaboration, is permitted as long as team success is given priority over individual success. Personal differences are set aside to achieve common goals. This type of teamwork results in a lasting mutual respect and camaraderie among the players.

The risks taken during the Apollo Program were "calculated risks", that is, they were clearly differentiated from "chance taking". Little significance in life has been accomplished without taking risks, be it personal, financial, or reputational. There is little progress without risk and the greater the risk, the greater the reward. This is particularly true in the case of personal, that is, life or death risk, in which decisions must be made more carefully and deliberately. In Project Apollo, crew safety was the prime consideration in design, development, test, evaluation, and spaceflight itself. The second priority was mission success. All risk-oriented decisions were made with these two priorities. There is a certain boldness and audacity required to "live on the edge", whether it is in life or death enterprises like military service and spaceflight, or in other circumstances where risk is a factor.

The personal qualities of the Apollo team were like those of most people who engage in innovative enterprises. They were highly individualistic, self-starters, risk-takers, and challenge-seekers. They were competitive, self-reliant, and confident in themselves and in their knowledge of the systems and operations of the Apollo rockets and spacecraft. They were comfortable to be either leaders or followers, depending on the circumstances. They also developed the capability to be brutally frank and honest with each other in assessing problems and solutions in contingency and emergency situations.

Moreover, the Apollo team was committed to excellence in all phases of development, testing, and flight. Their commitment was commensurate with the recognition that personal safety and human lives rested in their hands, not to mention, their personal pride in a job well done. Crew safety and mission success was their first thought in the morning and their last thought at night; a way of life in their profession. Such qualities can also be applied in less critical but demanding business challenges in terms of quality of products and services offered, individual and corporate performance, leadership, morale, training programs for users, and service to customers.

Finally, a positive attitude was a requirement for Project Apollo because first-time adventures are prime for miscalculation and ripe for frequent and rapid alterations to an otherwise well thought-out and vetted plan. One must be mentally prepared to respond to unexpected situations. A positive attitude includes preparing and working in the present as if we are currently living in a planned future outcome. When preparing to explore the Moon, for example, the Apollo team was trained to take action "when" they arrived at the Moon, not "if" they arrived. As the CAPCOM (air to ground controller) during Apollo 13 accident, I was asked what we would have done if the crew could not have been rescued. My answer was, "I don't know because it never entered my mind". That was the positive attitude of the entire Apollo flight team. Similar reactions surrounded repairs on the Skylab Space Station. Such positive attitudes can be applied in most of life's ordinary or extraordinary circumstances.

This book, "It WAS Rocket Science", by Tom Taormina, is a textbook about how to develop and provide useful products and services timely and economically by employing the well-defined and proven models used to plan and execute one of America's most complex and widely acclaimed engineering challenges. These same principles can be applied successfully at all levels of for-profit and non-profit endeavors providing products or services for the public or private sectors. Tom Taormina presents herein these techniques in clear and interesting detail. He should know; he was there!

Colonel Jack R. Lousma, USMC (Ret.)
Former NASA Astronaut
Pilot, Skylab 3
Commander, STS-3 (Columbia)

Foreword

Colonel Jack R. Lousma

VOLUME I
WE (SHOULD) LEARN FROM HISTORY

*"That men do not **learn** very much from the lessons of **history** is the most important of all the lessons of **history**."*
- Aldous Huxley

Space Shuttle Challenger Explosion, 1986

CHAPTER 1

HOUSTON, WE HAVE A PROBLEM!

What turned out to be the understatement of April 1970 was a wake-up call for those of us working on the flight of Apollo 13. Until the imperative "Houston, We Have a Problem" disrupted the customary air-to-ground chatter, we had been on a nominal flight plan for (what the press had dubbed) just another routine lunar space mission. Three astronauts were two days into a voyage from the earth to the moon when disaster struck.

I was having a cup of coffee, reading the mission log from the day shift and listening to the air-to-ground loop on the communications console. Those first words from astronaut Jack Sweigert caused me to sit up smartly in the chair, set aside the shift-log, and put a fresh light on my pipe.

I turned the volume up in time to hear the ground controller reply, "Say again, please?" Mission Commander Jim Lovell repeated the declaration, "Houston, it looks like we've had a problem." The astronauts reported they were rapidly losing power in Odyssey, the command module. There was also visual evidence that they were venting gasses into space. Either problem was potentially life threatening. Together, there were anomalies beyond those ever envisioned or simulated. We would learn that a required maintenance procedure performed by astronaut Jack Sweigert had precipitated an explosion in the service module of the spacecraft.

I immediately selected several more communications loops to monitor the dialogue between the various mission specialists. I leaned into the speaker hoping to follow the cacophony of voices and make some sense of what had happened. The plume of smoke from my pipe must have looked like the stack of a steam locomotive keeping time with my elevated pulse.

QUALITY ASSURANCE

The Quality Assurance (QA) Office was directly across the hall from the Second Floor Mission Operations Control Room (MOCR), Mission Operations Wing (MOW) in Building 30 of the Johnson Space Center. That MOCR was Mission Control for Apollo 13 (there was a duplicate control room on the third floor that was active in case of a systems failure on the second floor). I was a 25-year-old Quality Control Engineer assigned to the swing shift for this mission. There were six of us in the QA department of Philco, the prime contractor for Building 30. We staffed the QA office 24/7 during missions.

When no space missions were flying, my regular duties included inspecting the ground support hardware for defects, monitoring qualification tests, and performing capability audits of potential suppliers to NASA. During a mission, my responsibilities also included being part of the Technical Operations (Tech-ops) team. Our assignment was to be available in case there was a malfunction in any of the ground support hardware in Building 30. Should a problem occur, we were responsible for monitoring repairs, witnessing testing of the repairs, and re-sealing the hardware with tamper-proof devices. With the rigorous processes we used to build, inspect, and test the hardware before missions, there was seldom a need for us to do our jobs during a flight.

Since the Johnson Space Center began controlling manned space flight, QA's greatest contribution to lunar missions had been keeping the shift log current and staying close to a communications console in case we were needed. This was only my second opportunity to support a live mission, thus, I was more involved and living every minute of it.

WE DID WHAT WE WERE TRAINED TO DO

I became aware of a flurry of activity in the hallways outside the MOCR, but was engrossed in the communications chatter. The mission specialists were close to a diagnosis of the explosion and what the impact of the problem would be on safety and mission success. I followed the developing scenario that an oxygen tank had exploded causing the power loss in Odyssey. The Apollo spacecraft's used cryogenics, both for breathing and for creating electricity. This double-whammy had potentially dire consequences.

Never done or simulated, mission specialists shut down the command module, conserved the remaining power for reentry, and used the Lunar

Excursion Module (LEM) as a lifeboat. This power-down procedure was unprecedented. While they devised a rescue plan, the astronauts moved into the LEM as a makeshift lifeboat. By now, I had silenced the speaker and put on a headset for greater clarity. I was chewing a hole into my pipe stem and positioning myself emotionally to be part of the recovery team. Although I had no specific duties, I was ready to respond to a "call to action" and join the beehive of activity that was taking place on the communications loops and in the hallway.

All sense of time had vanished. Fortunately, the hours of simulation we endured as training for problems snapped me into reality. A familiar voice overrode the chatter on the communications console. "QA – Tech-ops, my loop." Tech-ops was the only functioning two-way circuit QA had and the volume was significantly above the drone of the other voices. Tech-ops coordinated all of the maintenance and configuration work on the ground support hardware and directed repair efforts during missions. "Tech-ops – QA – Go" was my trained reply. "Who is on duty?" the voice asked. "Taormina." was my only retort. In a rapid succession of two-way dialogue, it was determined that the only Polaroid camera in the building was in our office. "Get down to Sub-stores and bring all the film back to your office and stand by" was the final transmission before I relinquished my headset and laid my pipe down.

In the opposite corner of the second floor was a storeroom of operational supplies to support the needs of Mission Control. As I exited my office, for the first time, I became aware that the activity I had heard was the sound of programmers and mathematicians setting up schoolroom chairs in the hallway. There were at least a dozen of them, by this time, thumbing through computer printouts and manipulating slide rules. I had learned later that they were performing calculations to determine the best option for returning the spacecraft and the three astronauts safely to earth. By the end of the night, the rectangular hallway was crammed with a staff of specialists concocting various rescue procedures in real-time.

There must have been 20 rolls of film on the shelves in Sub-stores. I filled out the appropriate requisition form and hoofed it back to the office, film in hand. Almost immediately, a familiar face breached the doorway and asked, "Are you the QA guy?" As it turns out, he looked familiar because he was senior astronaut Frank Borman. Although we worked a few feet away from the MOCR, we almost never interacted with astronauts or flight controllers. Meeting Colonel Borman was a real treat for me. "Yes sir, that's me." was my only reply. He told me to get the film and the camera and follow him down the hall to a makeshift tactical control center. A crowd had gathered in a staff support room that was normally vacant during a mission.

Whether you recalled the televised events of those few days or you have seen it portrayed in Ron Howard's compelling (and highly accurate) movie, the rescue plan was hatched by astronauts, flight controllers, and mission specialists in a smoke-filled room on the second floor of Building 30. It turned out to be one of the most amazing demonstrations of teamwork ever. In mere hours, they had to contain the disaster and create a viable rescue plan. In parallel efforts, they would solve the issues of keeping the astronauts alive, returning them to earth, and making a successful reentry and landing in a spacecraft with little power and even less air to breathe.

One of the most accurate details in the movie was the chaotic moment in the tactical control center when the lamp in the overhead projector burned out. Rather than hunting down a new bulb, they reverted to chalk and blackboard to brainstorm the problems. My job was to take Polaroid pictures of the calculations and diagrams they were drawing on the board. After they were through with a particular scenario, I would wash the board and they would start over again. I turned over the photos to the appropriate specialists to convert the data into a revised mission plan.

I was the designated photographer for the rest of my shift. When I left the tactical control room, I noted my activities in the shift log, collected my pipe, and prepared to handover the special access badge to the next QA representative. The passing of the restricted-access badge (that gained us access to every corner of Building 30) was a ceremonial indication of a successful handover, mimicking the process the various teams of flight controllers conducted at the end of their shifts.

SPLASHDOWN AND LESSONS LEARNED

After the successful landing and recovery of the Apollo 13 crew, we had our usual series of splashdown parties, debriefings, and began reconfiguring the third floor MOCR for Apollo 14. Within Mission Control, we relived our respective roles during the recovery effort over coffee, but there was little sense of the impact those few days had on our lives, our careers, and our national history. After any event, life returned back to normal and I was soon back on the road performing capability audits of potential new suppliers to NASA.

I use the word "normal" cautiously these days because age and wisdom has caused me to be more appreciative of the pioneering work our team did in the sixties and seventies. Being young and idealistic can be synonymous with being oblivious to historical significance and the potential danger around us. Youth also includes a certain naïveté that masks the overwhelming odds against accomplishing seemingly impossible tasks, such as the rescue of Apollo 13.

Photo of the damage from the explosion in the Service Module of Apollo 13.
The extent of the damage was unknown until just before re-entry

For many years, I was relatively unaware of the personal and professional impact of my involvement and limited contribution towards that moment in history. Even though I was often asked what it was like to be in Mission Control during Apollo 13, they were answered with technical data, statistics, and process details. There may have been some ego woven into the answers (and certainly, the stories have become embellished over time), but I was mostly conveying to others that we just did what we were trained to do. After the adrenaline rush from the first hour of monitoring air-to-ground and flight controller conversations, the realities of my training became obvious when I went to Sub-stores and dutifully filled out a requisition form for Polaroid film. Although the problems were immediate and the situation was grave, we all did what we were trained to do. Why else would I have taken the time to fill out a requisition form in the midst of a crisis? The answer is that effective process procedures and training prevented anarchy and wasted effort.

We found tactical solutions to the problems at hand and implemented them the same way in which we found a strategic solution to the challenge from President John F. Kennedy that "…this nation should commit itself to achieving the goal, before this decade is out, of landing a man on the moon and returning him safely to the earth." I never knew who coined this statement, but I found it to be very accurate: "For centuries man has dreamed of traveling to the moon. We just made our minds up and did it."

CHAPTER 2
MY STORY

To borrow a line from Martin Sheen's character, Captain Willard, in *Apocalypse Now*, "There is no way to tell this story without telling my own." So let us start at the beginning.

I am not sure what gene I inherited that has relentlessly motivated me to be an innovator, leader, and to achieve successes beyond what should have been typical for an Italian-American kid from the streets of Brooklyn. Both of my grandfathers had come over on a boat from Sicily to Ellis Island. One became a baker and the other a zoo worker. My father sold pots and pans and then soda. We lived in a brownstone apartment near the elevated train. The highlight of my young life was the occasional visit to Ebbets Field to see the (real) Dodger's play baseball. We didn't have the money for tickets to the ball games, but the zoo workers and ticket takers were in the same union. So Grandpa, somehow, got us invited into the first-base seats by means of some barter system I was never privy to understand.

I was taught to be a follower by the nuns at my school and encouraged to smoke and be a juvenile delinquent by my peers. I did not have the stomach for petty crime and I only owned a brown (not black) leather jacket. In the social mores of the 1950's, that brown jacket labeled me as a non-player on the mean streets.

If I had followed my father's relentless career counseling, I would, today, be a retired electrical engineer, collecting a union pension, and spending my golden years watching baseball and war movies. My traditional rearing and inner city environment does not sow the fertile soil from which pioneers and innovators typically sprout, thus my theory is that I inherited a latent gene from one of my Sicilian ancestors who may have been an olive oil entrepreneur. This genome manifested itself in my passion for all things electrical, electronic, or wireless. I'm sure my parents kept an emergency fund just to replace glass fuses that I regularly caused to blow during my novitiate in electricity. Instead of graduating from puberty by surviving street gang fights, I survived electrocution.

The innovative gene must have sprouted along with body hair because I taught myself electronic basics, international radio regulations, and became a licensed ham operator at age 13. By age 14, I was building transmitters and erecting antennas. While my school chums were playing baseball, I was experimenting with how very-high frequency radio waves propagated through the troposphere. Instead of dating (lack of female companionship was not planned), I was working with older mentors, bouncing radio signals off the moon, and experimenting with newfangled transistors. I learned soldering instead of batting averages and drag racing. I may have been an early geek prototype. Instead of a pocket protector, however, I often wore a Civil Defense armband.

One of my mentors was an innovator in microwave radio and radio astronomy. He taught me that there were no boundaries to what is possible. I learned from him that failure in experimentation was a learning experience, while failure to try something new is a sin. He modeled for me a simple formula: that passion combined with a desire to achieve were the only prerequisites for attempting new ventures. By age 17, I was involved in my first entrepreneurial business building radio apparatus in another ham's basement.

JFK

Another seminal event transpired during my 17th year. In 1961, President Kennedy announced to the world that "this nation should commit itself to achieving the goal, before this decade is out. Of landing a man on the moon and returning him safely to earth." JFK was obviously speaking directly to me. I knew intuitively that I had been selected by fate to be a rocket scientist.

My family had always assumed I would become an electrical engineer. During my senior year in high school, my engineering predisposition and all of my non-traditional history and drive came together in the realization that my life's mission was, in fact, to help fulfill JFK's vision of dominating the space race. I tried it my father's way for a year at the State University of New York. Fortunately, JFK's call to duty was so compelling that, on my 19th birthday, I packed everything I owned into my Corvair and left the family home in New York for Houston. This move was pretty gutsy for someone who had only once ever been more than 150 miles from home.

President John F. Kennedy creating his vision

I enrolled in the engineering program at the University of Houston and somehow pestered enough people to land a job as an electronics technician with Philco, the prime contractor to Mission Control at NASA, Houston. I became part of the team that built the electronics that was being designed for the Mission Control Center and Project Apollo. I had a hand in building many of the green consoles we became accustomed to seeing on TV during the 60's and 70's and in the Apollo 13 movie by Ron Howard.

Before I finished my EE degree, I was influenced by another mentor to join the new discipline of Quality Control Engineering. Along with the many other spin-offs of the space program came new disciplines born of necessity and mine was based on developing innovative methods for preventing defects from happening. The stated mission of quality engineers is to ensure defects are not designed into products and services, rather than inspecting for defects after they were finished. To paraphrase an old TV commercial, we developed pioneering methods for "putting the quality in before the name goes on."

I'd come a long way - from twisting wires in grandpa's basement to being part of design teams that built electronic systems of brand new technology that were reliable enough to support manned spaceflight. Not only was I realizing my need for new challenges, I was thumbing my nose at conventional wisdom and traditional occupations. I was at home in Houston. The Dodgers relocated to Los Angeles and we were on our way to the moon by the time I was 20. All was well with the world, except for the Vietnam conflict and the Cold War. But that discussion can wait until a later chapter.

THE FOUNDATION OF THE APOLLO BUSINESS MODEL

For ten years, my quality engineering job took me to the far corners of North America to visit companies that were proposing to supply goods and services to NASA. My mission was to look at these organizations from the perspective of their ability to produce "quality" products. What I soon learned was that truly successful companies are a unique combination of visionary leaders, loyal and self-motivated employees, and a dedication to delivering true value to their customers. Over the span of 40 years, I have visited more than 600 companies and can report to you, without fear of contradiction that "quality" is a state of mind, not an activity of inspectors detecting defects after they happen. Quality is, in fact, about shared vision, leadership, and the pursuit of success through self-accountability.

As described earlier, one of my most profound life experiences was being on duty at Mission Control during the flight of Apollo 13. "Houston, we've had a problem!" The Apollo 13 team became a defining model for converting catastrophe into opportunity.

Those of us who worked on the rescue mission recalled the mandate from President Kennedy that included not only landing a man on the moon but also "returning him safely to the earth." We have well documented procedures. We were trained in those procedures and simulated them continually. We had clear roles, responsibilities, and contingency plans for potential disasters. Our leaders gave us clear direction, including Flight Director Gene Kranz' imperative, "Failure is not an option." We had the resources to execute our mission and the entrepreneurial authority to solve impossible problems. Most of all, we each assumed personal accountability for the successful recovery of Apollo 13. This experience was the foundation for my unending journey in embracing adversity, as a stepping-stone, to building a strong and united business culture where we all prosper.

The first thirty-five years of my life either afforded me a magnificent model of vision, hard work, and personal accountability as the foundations for long-term success, or it prepared me for a trip to Fantasy Island. The uncertainty began when I started applying my training and value system in corporate America. The term "it's not rocket science" took on a totally new meaning as I struggled to be a welcome contributor in a competitive businesses world[1].

Having fulfilled my first known mission in life, I left the space program in 1980. My next conquest was to apply what I had learned about quality at NASA towards companies that were actually competing to make money. Although I was a participant in the wasteful world of government contracts, I had also visited scores of for-profit organizations and was absolutely certain I had the formula for bringing the success of Project Apollo to American manufacturing companies.

I know; if you are not openly laughing right now, there is at least a snicker on your lips. My ambition smacks of naiveté, eh? I actually found a way to pull it off. In addition, my life is not as charmed as it may sound. I left out the parts about being drafted, getting married, having two kids, and being wiped out in two floods. I'll skip that character-building stuff because we all have similar tales to tell.

I DO believe that we create our own reality. So I found three different manufacturing companies that, over a 10-year period, afforded me the test bed to implement my process madness. I joined these companies as a quality control manager under the stipulation that within six months, manufacturing and quality control will be combined into one group under one leader (me). This was like suggesting that a fox guard the hen house in the 1980's. Wouldn't customers suffer without quality inspectors finding the defects built into their products? Wouldn't workers run amok knowing that there were no inspectors to hound them? In a traditional manufacturing operation, perhaps so. In a less conventional

---

1   I have since trademarked "It WAS Rocket Science."

environment where individuals are taught to be accountable for their own work and answer to each other for shielding the customer from defects, it can be quite a different story. No, I did not employ warm and fuzzy quality circles, nor did I run off the quality professionals. I made it all about personal accountability, enhancing the bottom line, and sharing success with everyone, but you will have to finish the book to find out the implementation details.

My experiments were with growing organizations that did not have many bad habits to break, so my successes only consumed seven-day weeks and my first marriage. More established companies, as I would find out, were less open to moving beyond chaos, command, control, and tribal knowledge. Thus, the concept of personal accountability was more difficult to implement. After 25 years of relentlessly building business processes around my NASA training, I still had part of my innocence intact.

FAREWELL TO CORPORATE AMERICA

My last job in corporate American was as the head of training and development at a division of Schlumberger, a multi-national conglomerate. My challenge was to transform a traditional manufacturing culture (of mostly female Hispanic and Asian workers) into one of Self-Directed Work Teams (SDWT). SDWT was one of the fads-du-jour of the late 1980's. Of course, all business fads (TQM, SDWT, Quality Circles, etc.) have their bases in modification of human behavior for the benefit of the shareholders. Thus, SDWT was destined to never work as advertised. Since it was close enough to my NASA-personal accountability model, I spent two years helping to create an environment in which folks could be self-motivated. My methods did not exactly follow the published texts of the day. In fact, the way we made it work was simply to listen to the needs of the employees and then, create the infrastructure in which they can be most productive. For example, my associate went to the leadership of the Houston Vietnamese community and enrolled in a class to learn conversational Vietnamese. She presented the SDWT concept to the community elders and was told emphatically that female Asians would never be able to grasp that concept because they were traditionally subservient and only worked when told what to do. It turns out that they actually became self-accountable when we listened to their need to have an area set aside for them to begin their traditional rice and fish cooking each morning (before work) for their noon meal. That "motivation" led to an immediate decrease in defects and a great aroma in the plant. I undertook SDWT with the Hispanic group, who would much rather be told what to do (according to their cultural elders) and then be punished for their mistakes. By talking with these folks, we found out that their biggest motivators were their children getting into trouble while the parents were at work and the financial chaos most of them suffered because their spouses were losing their jobs in the declining petrochemical industry. Being part of a Self-Directed Work Team was clearly not on their radar screens either. We located an inexpensive day care facility close to the plant, gave them access to family counseling, helped their spouses find outplacement help, and had the credit union come in and teach classes on household money management (during lunch hours). Once again, defects fell and productivity increased because quality inspection had been replaced by personal accountability by individuals who were treated with some dignity and respect.

The results of my incubator was about to be rolled out to the global corporate leadership when the parent company in Europe decided to outsource all manufacturing and furloughed almost all of the assembly workers. I was "fortunate" to witness the birth of the new phenomena of corporate "outsourcing" to avoid dealing directly with human beings and to cheapen products until they were no longer safe and reliable.

THE ENTREPRENEUR

Although all three of our sons were in college at the time, we had a family meeting and I was encouraged to leave the uncertainties of corporate America to pursue the next logical step in my evolution, independent consulting, and training. With all I had to offer, surely I would be in demand as a visionary, pioneer of quality, and modeler of human accountability.

You are laughing again! I can hear your pragmatic assessment of my plans as being (what was the word I used earlier?) naïve and, perhaps, unrealistic. Who am I to be telling corporate America how a business should be run? Where are my Masters in Business Administration and my PhD's in Corporate Psychology? I did not have the public visibility of Tom Peters and Anthony Robbins, nor did I see the future (apparently) as clearly as Joel Barker did. I didn't have a clever program like Hammer and Champy's *Reinventing the Corporation*. All I had were my street smarts, an incredible 14-year education at NASA, and ten years of helping oilfield service companies grow the efficiency of their manufacturing operations. Oops, I forgot to mention that the 1980 to 1990 decade saw the decline and fall of the petrochemical business and the lack of customers dragged several of my success stories into inevitable bankruptcy. Maybe that's why I was not being offered a contract for success-story videos like Tom Peters. What the heck. I know that I have the formula for the future success of American businesses in my model of (1) a shared vision led by visionaries creating a work environment that encourages personal accountability and (2) everyone sharing the rewards of success. What I had to offer was so self-evident and so much on the remedy for thriving on chaos that I could not fail. Damn the torpedoes; full speed ahead. My new bride and I will become fearless entrepreneurs who hoist the flag of quality and artisanship and begin a crusade to rid the country of corporate inefficiency (enough laughing, already).

For those of you who haven't had the opportunity to bid farewell to corporate America and venture out into the world of consulting (with the goal of sharing your passionate vision with eager business leaders), the experience is sort of like being permanently unemployed while dealing with rejection on an hourly basis. The success story of Harlan Sanders and his Kentucky Fried Chicken had its austere beginnings with the Colonel visiting over 100 entrepreneurs before someone finally saw the possibility in KFC. There were days when I felt like I was living in the back of his old Cadillac, getting ready to mix up my 17 secret herbs and spices for another closed-minded business owner. Fortunately, one of my earlier duties as manager of training and development was to learn how to implement the new international quality standard, ISO 9000. For those of you not familiar with ISO 9000, the European Union commissioned a quality standard that was universal in nature that can be followed by any business and culture within the EU. It was so elegantly simple in its tenets and clarity, that, since 1987, it has been adopted in more than 140 countries. I embraced ISO 9000 as a baseline platform from which my business process methods can be implemented and was able to find a market implementing ISO 9000 in petrochemical service companies. The American Productivity and Quality Center in Houston commissioned me to develop their ISO 9000 training courses. As my practice expanded to other industries, Productivity, Inc., contracted me to present ISO 9000 seminars that related practical experiences for effective implementation of the Standard. By design, my courses were peppered with my lust for decentralizing quality control and causing individuals to become more accountable for their own actions. For those of you who have been through ISO 9000 implementations that were ineffective or overhead expenses, it really can be a platform for business process improvement; not more corporate bureaucracy.

I started writing books on ISO 9000 as it related to business process improvement and effective leadership (10 to date) and I became a regular contributor to quality assurance and quality management journals internationally. I coined the term "ISO 9000 as a Profit Center" in an attempt to highlight my proven methods for causing quality management to contribute to the bottom line, instead of being an overhead cost center.

THE GLORY DAYS OF DELL COMPUTER

In 1999, I finally got my dream consulting assignment. A visionary quality manager and divisional VP at Dell Computer[2] called me to help bring a unit of Dell into ISO 9002 compliance. This was a splinter group that was not as structured as the other divisions of Dell and needed critical resuscitation to get their quality system up to par. Again, I managed to concoct a contract that had some non-traditional stipulations. I would get the division ready for an audit in two months, if we would use the draft of the 2000 revision of ISO 9001 as a model for the rest of Dell to take their ISO 9000 certification to new levels of business enhancement (as opposed to just basic compliance). The 2000 revision was a rewrite of the Standard that not only documented a minimal quality system for companies to model, but it set in place requirements for companies to sign up for "continual process improvement" as an integral underpinning of their business operations. The new revision not only embraced many of my NASA-derived methodologies, but also made them a requirement for companies that would upgrade to the new Standard. I was a happy camper.

The system we implemented at Dell started a yearlong evolution from an ISO Quality Management System (QMS) to a Business Management

---

2   My team and I conducted a total of three major consulting assignments for Dell. We did an evaluation of Dell University and helped them transition to on-line learning. We helped write their Supplier Quality Manual which was integral to their revolutionary supply chain management program. At the end, four of the group went to work for Dell full time and The Virtual Group disbanded.

System (BMS) to a totally paperless system called Business Management Interactive System (BMIS). BMIS became one element of Dell's current system called Global Business Process Improvement, which is a mandatory part of every manager's evaluation and compensation system. To date GPBI has contributed multi-billion dollars of auditable savings to the Dell bottom line. This case study is documented in my book, ***Implementing ISO 9001:2000 – The Journey from Conformance to Performance*** (Prentice Hall, ©2002). Unfortunately, the book was published 3 weeks after 9-11-2001, so you won't find it on any best-seller lists.

I took the Dell story on the road to quality conferences around the country. Surely my message would become the model by which visionary business leaders would flock to embrace the tenets of Quality as a Profit Center. While the presentations were well received by my peers and resulted in several more book contracts, I was still unable to get my message across to business leaders (they don't typically attend quality conventions). I apparently had "Quality Geek" tattooed on my forehead and when I gave private presentations or talked to professional societies, I seldom got any takers for test-driving my methods. I was, again, feeling very much like Colonel Sanders after approaching his 50th investor. With 14 years of consulting, I had enough successes to make a living, but I'm still frustrated about why enlightened business leaders are not jumping at the chance to dismantle costly quality control organizations and embracing to empower their work force to perform splendidly as individuals and as contributors to a common vision.

THE FS 9000 FOLLY

The fact that Dell kept us busy with new projects beyond BMIS encouraged my business partner and I to invest four years of our time (and much of his retirement funds) to present our cost savings approach to quality to the organizations that dealt with money as their main product. With the network of contacts provided by my business partner (a retired Merrill Lynch exec), we presented the ISO 9000 as a Profit Center concept to banks, insurance companies, and securities firms as a breakthrough in minimizing financial risk. We coined FS 9000 and formed a not-for-profit organization that was destined (we hoped) to blaze a trail through the international financial community. Several visionary organizations in the banking and auditing businesses encouraged us to press forward. We hosted several conferences in New York City and in Boston. We had recruited founders and were ready to roll out the project when 9-11-2001 happened. Of course, FS 9000 was placed on-hold as the world's financial community regrouped and reverted to survival tactics. We did have one ray of hope. The Federal Reserve Bank of New York was quoted in saying that "If it weren't for the discipline we learned from ISO 9000, the Fed Bank of NY might not have survived 9-11." After the smoke and ashes cleared, we revived FS 9000 and were once again at launch stage when the Sarbanes-Oxley (SOX) act became law. SOX, as it is called today, created a new set of quality control measures in which corporations must implement to avoid Enron-type scandals from being possible in the future. Instead of embracing FS 9000 as a platform in which to implement SOX, the supporters of FS 9000 dropped by the wayside to open new departments of SOX compliance. Depression set in (briefly) as we counted the money and years we had poured into the futile vision of proactive quality management being embraced as a viable business model in the financial world. Was there no company besides Dell Computer that had the vision to embrace our methods? Were my 17 herbs and spices destined to always be a family secret?

PROVING MY THEOREMS

Thanks for hanging in through my life-journey thus far. You may be seeing the thread developing that will lay the foundation for the book title, *It WAS Rocket Science*. We will be quickly tying 35 years of history together with present-day reality to put the cornerstone into the foundation of my Apollo Business Model thesis.

Since moving from Texas to Northern Nevada in 1997, I was again privileged to land in a target-rich business environment of growing companies outwardly eager to realize their greatest potential. As I began introducing myself (through seminars and workshops) to the business community, I was shocked to find that the business leaders who stated that they wanted to grow their companies believed that they had to do so with a work force that was apathetic, unmotivated, disloyal, and untrustworthy. Most of these otherwise-insightful folks accepted the fact that an internal militia was necessary to ensure that the workers produce a reasonably reliable product or service. This perception was vividly modeled for us in Nevada by the gaming industry. Casinos and gaming equipment suppliers are mired in regulatory constraints. As a result, industry leaders have mastered the technique "management by absolute control" for every aspect of human endeavor. Prescriptive regulations (which includes jail time for noncompliance) combined with "money" as the product leaves little incentive for experimentation with best-in-class proactive quality tools. With the ongoing construction boom and minimal unemployment, most non-gaming business leaders are convinced by their gaming colleagues that their hourly workforce can never be trusted without highly visible (and expensive) command and control procedures. They also believe that their exempt employees will flee the moment more money was offered to them, thus, they invest no effort in ensuring employee loyalty. Strictly enforcing quality controls and continuing new employee trainings are just a normal cost of doing business in Nevada, or so it would seem.

The other major industries in this area are mining and warehousing. Neither of them have the need to be certified to ISO 9000, so there is little opportunity for me to segue into Quality as a Profit Center. Still, I have found colleges and outreach programs to be receptive to raising the bar in business process effectiveness and I have had a number of successes locally.

I have also assembled a formidable consortium of subject matter experts into my consulting inner-circle. As we began working together and compared notes about our recent assignments, we were collectively shocked by the number of "Non-Failures" we encountered. A Non-Failure is exemplified by a company that hires us to apply some sort of Band-Aid to help them through a crisis while never (subconsciously) intending to cure the underlying causes. In other words, business leaders were looking for a magic pill that would relieve their immediate symptoms without investing in the process of finding a cure. They are not lazy executives, so there has to be a common theme by which they choose to shop exclusively in the "quick remedy" aisle of the drug store.

The second category of companies we've collectively identified are those that outwardly have all the appearances of being healthy and productive, but there is much turmoil behind the polished public image. I have labeled these companies as "Apparent Successes" because, under the façade, they have systemic issues that are potential time bombs for disaster. The leaders of these companies are outwardly open to business process improvement tools and to raising the maturity of their workforce. Once we began working with them, however, their managers and seasoned employees quickly admitted to the lack of clear direction, management by crisis, favoritism, or other counterproductive leadership traits that stifled personal creativity and accountability. Once the business leader discovers that we have uncovered the issues that require him or her to modify their behavior, the consulting assignment is usually terminated.

There is an even more bizarre third category of companies we have encountered recently that are consumed with fatal flaws. Yet they stay in business for years, far beyond what fiscal and rational business logic would dictate. We call these operations "Terminally Unsuccessful."

My colleagues and I have consumed many man-months conducting analyses on case studies, seminars, workshops, and training programs in hope of discovering a common thread for what causes companies to be anywhere from apathetic to absolutely closed to fixing problems at their root cause. More specifically, we were really challenged to explain why business leaders allow dysfunction, turmoil, employee apathy, and lack of accountability to exist within their organizations as an inevitable reality of being in business.

We have objectively and emotionally sifted through our data looking for common threads of success and failure. Our research has been with businesses in representative industry segments from manufacturing and construction to retail and service industries. Our brain trust[3] is not attempting to compete with the great business schools that are all searching for the secret of business success. We are, instead, utilizing the data collected from our respective successes and failures to discover why most businesses unconsciously condone and permit an environment of mediocrity to flourish in their companies. In fact, we have discovered that many companies spend significant energy on perfecting the mediocrity that exists in their organizations. We have been able to represent our findings with an equation we call the 4P model.

People + Process + Power = Performance

The model states that the Performance of a company is the is the sum of the productivity of the People, the effectiveness of the business Process, and the leadership provided by the Power (the person who actually runs the company). By using the mathematical metaphor, the effectiveness of any one of the three factors can be seriously degraded by poor results in either or both of the other two factors. We have found, unfortunately, that books and trainings, which deal with tools for fixing processes (including the current wave of Six Sigma books, belts, and certifications), ignore the other two key ingredients in the formula.

---

3 http://comstockpride.com

The next most popular panaceas are those that deal with modifying the behavior of people in order to achieve a "team" of highly motivated worker bees. Behavior modification has never worked and never will work. Thus, fixing processes has only limited success when apathetic workers are running them. The arena in which almost no consultant or facilitator dares to tread is into the inner sanctum of organizations where the true power resides. The principal reason we avoid dealing with senior management is that business leaders seldom have any perception about how the power they wield affects those who work for them and how their power is perceived by their inner circle. Consultants who attempt to break through an entrepreneurial leader's cloak of power are often met with "I've been successful thus far. Why would I want to change now?" These are the folks who buy the business-triumph stories lining the bookshelves and somehow believe that by studying Michael Dell's success story, they will, by osmosis, become more robust leaders. They watch CNN and blame their own corporate woes on governmental failures rather than on the apathetic nonperformance of private businesses. These so-called business leaders should instead be watching National Geographic or the Discovery Channel to witness the stories about the new Shanghai, Chunking, Taiwan, Dubai, and other emerging centers of world power that are positioning themselves to convert our American businesses and social infrastructures into a third-world has-been.

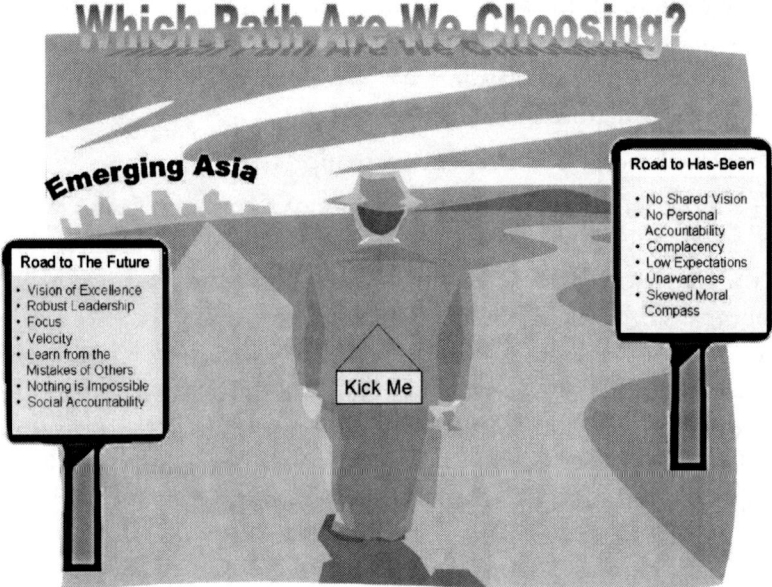

The rest of this book will deal with how we can stop squandering the lessons of excellence that built this great country and how we can turn around the current national obsession with perfecting mediocrity. It will highlight the lessons learned from Project Apollo, 9-11, and Enron as compelling models for us to turn evolving disaster into a renewed mission for America to rededicate itself to being the role model of excellence for the world. Instead we are growing obese on our perceived successes while the center of business success moves westward across the Pacific Ocean.

CHAPTER 3
THE RISE AND FALL OF THE AMERICAN BUSINESS EMPIRE

This chapter will introduce you to a few of my foundational theorems based on personal and anecdotal experiences.

YOU MUST MAKE A FRIEND OF HORROR

While I did considerable research for this book, much of the foundation for It WAS Rocket Science comes from a lifetime of personal and professional experiences, often rich in painful lessons. Since my nature is to make lemonade from lemons, I try to learn from each experience and not repeat any expensive faux pas. In concert with learning from family and business mistakes, 40+ years of working as a quality engineer has prepared me to almost instinctively scrutinize and catalog the details of any meaningful event. My training has included analyzing the captured data, converting it to knowledge, and then codifying the knowledge. I wrote my first ten books in an attempt to share the resulting experiential and anecdotal knowledge with others, so that they might avoid pitfalls that I paid dearly to experience.

In addition to my personal incidents, I have also been exposed to a treasure trove of events from working with more than 600 companies. These experiences have ranged from ridiculous to sublime. My engineering mind has plotted a "trend chart" of these observations. The data analyses indicate that, at this point in time, we have passed "absurd" and are moving towards "ridiculous." We are on our way to the extinction of personal accountability, artisanship, and customer service on our journey toward perfecting mediocrity.

For a glimpse into my experiential database, I have picked a brief selection of ridiculous-to-sublime examples that form some of the data points on my chart. It is from this exemplary information that I will present to you my earliest symptoms of our current national obsession with perfecting mediocrity.

I begin this journey with another quote from *Apocalypse Now* – Marlon Brando's soliloquy "Horror has a face... and you must make a friend of horror." The character, Colonel Kurtz, was referring to the horrors of war. I am borrowing this metaphor and applying it to the horrors of becoming victims of our own success. Until we face the realities of our societal erosion, we cannot reverse the trend. In later chapters I will share some of my inspirational experiences, in hopes of reversing the direction from decline to growth. To start, however, I am compelled to slowly begin by building the scenario for the foundation of our ethical decay by relating a number of every-day personal experiences, each rich in examples of behavioral absurdity.

THE DRYER

On an otherwise uneventful Saturday morning, our five-year-old gas clothing dryer quit drying clothes. The device would start correctly, but the gas would quickly shut itself off. As Mr. Handyman, I checked all of the obvious possible symptoms such as the plug not being in the wall and the propane tank being empty, but I quickly resigned to call the local authorized service agency.

We have used this company a couple of times. Other than the repairman appearing to be a retread biker and a heavy smoker, his ability to diagnose and to fix washer/dryer problems had previously been adequate. This time, however, the person who showed up (toolbox in hand) was a 19-year-old who gave us the first impression of having stayed up all night watching horror movies. He would not look at me directly and spoke under his breath, indicating an immediate minus-two on my scale of warm and fuzzy feelings.

He removed the usual clutter from the top of the dryer and spent the next five minutes poking and prodding. With a neutral expression on his face, he proclaimed that (sic) my guess is that the problem was the thermostat. I wasn't sure if this was the diagnosis or an invitation to play 20 questions.

After punching on his cell phone for a few minutes, he asked if I had a phone. Since my annoyance meter was already activated and registering at quarter scale, I bypassed the chance to be sarcastic about "having a phone" and pointed him to the landline. After a few minutes of dialogue with headquarters, he announced that the thermostat was $79.00 and a special order part. With two service calls, the total bill would be just under $200.00. I agreed to pay for the first service call while they ordered the part as I evaluated my options. He could not find his credit card machine or his pen, so he departed empty handed, leaving the clutter from the dryer on the utility room floor and a puddle of oil on the driveway.

Not really expecting a follow-up phone call, my wife contacted the company to get an estimated date for the repair. She was told that the part should arrive in a few days and that she would be back in warm air by the weekend. Ten days later and two more phone calls, the part still had not arrived. By this time she had spent two of her precious Saturday's at the Laundromat and her annoyance meter was already at the top of the scale. The young fellow returned on a Monday evening, installed the part, proclaimed victory, borrowed a pen to fill out the credit card voucher, and, again, left a clutter on the floor and a second puddle of oil in the driveway. On Tuesday morning, the very same symptoms of cold air continued to delay the drying process. Another call to headquarters (by now memorized in speed-dial) resulted in an apology and a promised return visit no later than Wednesday morning. About one PM on Wednesday, I punched speed-dial and reminded headquarters that it was no longer morning. The "manager" assured me that I misunderstood the previous conversation and our appointment was scheduled between two and four PM. I seized the opportunity to calmly relate my dissatisfaction with the service to date enumerating my specific concerns about the skill set of the service technician. I was assured that if this visit wasn't successful, the next step would be to send out the "senior" technician, who was the young man's father (the retread biker). The manager also assured me that this service call would be at no charge. Being comforted by the free service call and the thought of wasting another day on this project, I waited for junior to show up.

On arrival, he ceremoniously removed the same stuff from the dryer, opened the top, and stared endlessly at the rotating drum and the dark inner-workings of the dryer. I offered him a flashlight and then walked outside to check the exhaust air. I returned and reported to him that the air was still cold. Startled, he spoke his first words, proclaiming that he just didn't understand why it wasn't working. He asked to borrow a roll of electrical tape. I watched him cut some wires and return to watch the drum rotate. I asked what he had done. He told me he had cut out the high-temperature thermostat. I asked if that wasn't in the circuit for the purpose of preventing a fire. He didn't answer the question, but told me that it didn't help anyway. After splicing the wires, he proclaimed that "his best guess" was to replace the gas solenoid actuators, which he "did have on the truck." After five minutes of clanking, I again heard the drum rotating. I saw the first smile on his face. He again proclaimed victory. His success speech was something to the effect that the actuators probably failed after he left the first time and I should just pay him $36.00 for the parts and call it even. Speaking very slowly and deliberately, I reminded him that the symptom was exactly the same after installing the $79.00 thermostat which was probably not defective and that I was not paying another cent until the machine had operated successfully for at least one day. He borrowed my pen again, wrote up the service ticket, and left without further comment. Again, he left behind a clutter on the utility room floor and a third oil spot on the driveway.

THE HEART ATTACK

I've been dealt two different hands in life's genetic poker game. The first is that I am, by nature, healthy and robust. Aside from a tonsillectomy that I don't remember, I had never been hospitalized until I was just shy of my 60th birthday. The other hand I drew is that of constantly fighting my body's desire to be heavier than the insurance companies suggest - more than I would ever want to be. There is a lifelong history of unbending intent in discovering the metabolic contributors to this dilemma, but that's another story.

We live at 6,500' altitude. In the eight years we've been here, I had symptoms of altitude sickness, but the incredible beauty and pristine environment made the inconveniences worthwhile. Recently, however, I began experiencing shortness of breath with only minor physical exertion. My treadmill stamina has gone from half an hour to about ten minutes.

My longevity objective is to be shot by a jealous husband at age 95. The allegation would be false, but I would be so proud for being accused that I would die smiling. With that said, I have at least another 30 years of planned professional and personal objectives still in front of me and as advised by our family doctor, I scheduled visits to see a cardiologist and a pulmonary specialist. With modern medicine being at an advanced 21st century-state, I was confident that a clear and exact diagnosis was at the end of a few simple tests.

The cardiologist was an affable chap who listened to my symptoms and history intently asking the appropriate questions. He prescribed an echo sound of my heart to explore the possibility of visible coronary issues. This non-invasive procedure sounded like a logical first step.

The technician who performed the test grunted and pushed buttons ceremoniously until she had collected the desired data. I asked if there were any visible issues that she could discuss. I was told that the protocol was to have the doctor read the results before offering any observations. I immediately translated that story into a mechanism for maximizing billable hours by all concerned and minimizing rising insurance premiums by having a technician only say, "Here's a towel to clean that gel off your chest."

I seldom see a doctor more than twice a year for checkups, but I was eagerly awaiting my fourth appointment to get the results of the echo sound. My confidence in modern science was validated by the cardiologist's report that my left ventricle appeared to be pumping only 80% of its normal flow, which he explained, was the likely cause of my shortness of breath. He went on to describe how some previous illness might have damaged the heart or there might be coronary disease that was not obvious on the echo sound. The next step in the process would be to have an angiogram to take a peek inside the coronary arteries for any signs of restrictions.

For a short moment, I was flush with the anxiety of having a TV camera launched into my groin and then spending the night in the hospital. I got over it after talking with trusted friends who characterized the procedure as totally uneventful.

The process itself was an outstanding experience in customer service, from the admitting group, to the preparation, and through the procedure. The catheterization lab techs even asked what music I wanted in the background during the 45 minute operation. I selected the Eagles and they found the appropriate CD around the time I was being injected with happy juice (I must remember to drop Joe Walsh an email and let him know that his vocalization of "Life's been good to me so Far." was resonating in the background as the catheters were being injected in the foreground).

The satisfactory experience came to an end when I was placed in a post-op room with three other men. Two of us had rather mundane procedures and were there for a night of observation. The other two were critically ill and in serious discomfort. The ward nurse was Nurse Cratchet's understudy. She and her merry band of orderly's were as attentive and responsive to our needs as a New York City cabbie would be to a fare on a rainy day. I have never seen such rampant incompetence in an environment where life and death are only a short corridor away. I am not going to go into this excursion of gratuitous mediocrity because I still haven't decided if I'm going to file a formal complaint, so we'll move on to the more wanton examples of perfecting mediocrity in modern medicine.

After observing a night of continual neglect of our two seriously infirmed roommates, I was pleasantly surprised by an early morning visit from my cardiology team to discuss the results of the procedure. They explained how they had placed a stint in one artery that showed some signs of blockage, but I had a clean bill of health in the remaining cardiovascular system and in the heart muscle. After receiving my post-op instructions, I was sent home with a request to come to the office for a two-month follow-up.

During those two months, I was amazed at the almost non-existent side effects of the procedure. I was also proud of myself for being free of the cardiac disease that plagues many of my contemporaries and of my total cholesterol that stayed around 140. I was actually eager to be complimented again by the doctor and was expecting that shortness of breath would be an ongoing issue because of the restricted flow on the left ventricle. The prognosis would have some cool name like aortic stenosis.

The same doctor saw me for my two-month checkup. At least it looked like him. I even looked to see if he was him, who was reading my chart after his opening greeting, "Well, we sure don't want you to have another heart attack." I caught myself automatically asking, "What heart attack?" After fumbling through my chart, he changed his story to, "Well we caught the rampant heart disease just before it would have led to a serious heart attack." Again, an involuntary response flowed from my lips as I urgently queried, "What heart disease?" Again, he paged through the chart for what seemed like hours. "You have serious coronary disease and we have to get your cholesterol under control, young man." Since he was about the same age as my youngest son, I felt patronized by a person who had not taken the time to read what he had written in my chart two months earlier, before starting his "Autumn of my life" speech. After that remark, I said "Doctor, my total cholesterol is 140. How low would you like it to be? My blood pressure routinely runs 110 over 70. How much lower should it be?"

Obviously flustered at my challenges, he went back to the safety of reading my chart while composing his next offering. "I think you should accept that you have heart disease. It is the product of our modern lifestyle. Here are some prescriptions for you to take. Come back and see me in six months and we'll reassess the situation." Even more perturbed by the automatic dispensing of pills for a disease I didn't know I had (or do I?), I asked if either prescriptions would help the symptoms of shortness of breath. Confounded by my question, he went back to consult the chart again. "Oh, the angiogram did not show any ventricular problems. Your heart is fine." I then related my visits to the pulmonary doctor where their battery of tests gave my lungs a clean bill of health and asked the inevitable question, "After four months and over $65,000 worth of tests and procedures, what, then, is the cause of my shortness of breath?" He closed the chart, removed his reading glasses, looked directly at me, and said "You just need to lose some weight and consider moving to sea level."

THE BIG WHITE SUV

As discussed in the last vignette, we are blessed to live in one of the most beautiful spots in North America. We have a mountain-top home nestled in ten acres of pinion pine and juniper trees. My wife and I both drive mid-sized SUV's. They are the vehicles of choice because of the all-wheel-drive and their utilitarian capabilities. We owned five of the same model over the course of time and were very pleased with their ability to handle the winter driving challenges, especially our 1,000' long serpentine driveway with a 5% uphill grade. The previous owner-builder must have had a preoccupation with mountain roads because he built the driveway along the most indirect and difficult traverse imaginable, from the road to the house. Aside from neglecting to make it wide enough for a full-size fire truck or moving van to negotiate, he also added hairpin turns that made snow plowing a journey through a rat's maze. We also had to pay $23,000 for a proper asphalt-surfacing job after a UPS truck got stuck several times during our first winter on the mountain, but I digress.

For several years, the all-wheel drive in our favorite SUV model had always served us well, negotiating the mountain roads and our driveway, even in the most challenging snow storms. In late 2001, we were offered some extremely tempting customer loyalty incentives and traded in for the 2002 Big White SUV. I dubbed it "big" because the very first issue we discovered was that the body styling had changed from the previous models and it would not fit into the niche I had carved for its predecessor in my garage. Annoying and strike one. The next challenge was that it was difficult to start on cold mornings. A trip to the dealer yielded no explanations or resolutions because it started just fine at the dealership, 2,300' lower in altitude. After a number of visits to the dealer, my favorite service writer shared an anecdote with me that several of us may have uncovered a design issue with the fuel injection system. This flaw manifests itself with difficultly starting at high altitudes. He was going to consult with the manufacturer and report back with a cure. Weeks and a number of inquiries later, his report was offered "off the record." Indeed we had discovered a design anomaly. Since there were very few of these models that resided at high altitudes, the manufacturer did not see this problem as critical, nor was it at the top of their list of priorities to fix. His advice was to "Live with it." Strike two.

Two weeks after taking delivery of Big White, we had our first snowstorm. As we negotiated the seven-mile grade from town to home, the snowfall increased dramatically and the ground was covered with about two inches as we approached our driveway. Being accustomed to the handling of the previous SUV and knowing the driveway, we started the 1/5-mile climb as we had many times in past winters. Within about a hundred feet, the vehicle went out of control and we came to a stop some distance from the driveway lodged in a patch of sage brush. It was snowing smartly and we really had no reference as to where we were. Fortunately, we had been given a year of "free" access to the roadside assistance feature built into the vehicle. My wife pushed the big red emergency button and was soon connected with a voice of compassion and potential rescue. After being read the stock speech of comforting reassurance, the representative asked if we were on a public road or on private property. Her GPS plot of our location indicated that we were at least a mile from the main public road. We confirmed her suspicions that we lived on a private road and our driveway was some distance from the main drag. In a very patronizing tone, she reported that they did not provide service on private property. After a minute or two of debating, we were told that we should have read the details of the service agreement, which clearly states that they only provide "roadside" assistance.

Since the snow was unexpected, we only had light jackets and dress shoes. We felt like the Donner party making our way from the abandoned vehicle to the house in a white-out blizzard.

Later that day, the snow stopped and one of our thoughtful neighbors came to our rescue, towing out Big White with a winch truck. The roads were too warm for the snow to stick, so we made an immediate bee line to the dealership and demanded the use of another vehicle while they figured out why the all-wheel drive did not function. The next morning we got a call from our cheerful service advisor notifying us that we could come and pick up Big White. He wasn't sure what the problem was, but we could talk with the service person when we got there.

Mr. Goodwrench was an introverted sort, who likely avoided contact with the customers as much as possible. After some probing, he offered the following explanation. On the 2002 model, the all-wheel drive was not mechanically engaged as it was in previous models. It was now selected on or off by a computer command, that had not been engaged at the factory. We were reassured by the fact that it just took a few minutes to remedy the oversight. His response was mind-numbing and strike three.

My wife and I had a different take on the word "oversight." We had several conversations with the chain of command at the dealership where we related the scenario of how life-threatening the "oversight" could have been when I did not have the use of the all-wheel drive I expected on a mountain road in a snow storm. While there was measured indignation on the part of senior management at the manufacturer and reassurance that this would never happen again, neither the dealer nor the manufacturer every acknowledged that this event was potentially catastrophic and offered any explanation as to why the roadside assistance organization refused to help us when we were in distress.

A year later, the all-wheel drive system again malfunctioned in the first snow storm. One more trip to the dealer revealed another internal factory bulletin that should cure the new (and unstated) problem. It didn't. We demanded that they take the vehicle back because of the cold weather starting issues and safety problems with the all-wheel-drive. We were told that they couldn't do so because we waited a year to notify them of the problems. We reminded them that we did notify them a year earlier and that the all-wheel drive was only a necessity in the winter and winter driving conditions only occurred during the winter. The manufacturer declined to ever acknowledge the problems to us. The dealer "allowed" us to trade Big White for a newer model of another brand that used the same body styling and drive train as our previous vehicles.

DRIVING SCHOOL

Returning from a trip to Portland, OR, a representative of the California Highway Patrol took issue with my need for speed and we exchanged pleasantries somewhere in Trinity County. An option offered was to complete driving school and have the ticket expunged from my record. Since I am a realist, I acknowledged my infraction and took advantage of this opportunity to get a discount on my insurance by completing driving school. On a crisp October morning, I drove an hour to Truckee, CA, for 400 minutes of obligatory training. Little did I know that this day would be rich in data to add to my study of mediocrity.

The instructor was a sociable chap and managed to consume the first ninety minutes by going around the room and interviewing the 21 attendees. It wasn't long before I was scribbling notes as each perpetrator told their tale of traffic infractions. I'll share only a few of the more poignant stories with you because they leaped out at me, as examples of how our neighbors' value systems are manifested in the context of driving a vehicle.

Ms. W was a clinical psychologist. She explained the circumstances of her speeding ticket for going 80 in a 65-mile zone. "My brother had committed suicide the day before. I was in a state of total denial and do not remember anything about driving that day, nor do I remember speeding. I guess I must have been driving erratically because of my state of mind, but I don't remember."

Ms. X was 82-years-old and was returning home from a golf tournament when she was stopped for 70 in a 55. Her explanation was that she was talking with the other folks in the car and there is no sense of speed in her E class Mercedes.

Ms. Y was 18-years-old. Her citation was received before she turned 18. She was out for a ride with other underage drivers and was cited for violating the statute that prohibits underage drivers from carrying other underage passengers without supervision. The part of her story that is worthy of noting is that she graduated from high school and living at home until she decides "if she, maybe, wants to go to college somewhere." She has no job, no real interests, and her parents paid the fine and traffic school fees. She thought the ticket was unfair. She saw nothing wrong with breaking the law because Mom's Escalade is very safe and Mom did not care if she took her friends out for shopping trips.

Ms. Z was 16-years-old and had been driving for 2 months. She also had a car load of other children in mom's new GMC Yukon and was headed out to a party after the 11PM curfew for underage drivers. Her interview with the instructor was extensive, as she related the details of her undisciplined adolescent life. When she got the ticket, mom did not comment and dad thought it was funny because she got cited only 2 months after getting her license. Mom and dad both knew that she was out with friends after curfew, but (allegedly) saw nothing wrong with a 16-year-old driving over the state line into Nevada, on a weekday night, after 11PM. Her parents pay for the gas and insurance and have not revoked her driving privileges. Part of her fine was community service for which she was to fill two trash bags along the roadside and return them to court. She went home, filled the trash bags with household trash, returned to court with the trash, and was smug with how clever she was for faking out the judge.

Ms. A was a school bus driver. She was cited for speeding in her personal vehicle. She only worries about her speed when driving the school bus. To quote "The cops can catch me every day on that same stretch of highway." Her values include too much attention being paid to minor traffic violations and too many non-violent people in prison.

Mr. B was 68 and was taking his girlfriend and her twin sister on a trip to Southern California. According to his recollection of the infraction, he was so distracted by having the twins in the car that he was not paying attention to his speed.

Mr. C is 65 and on disability from three heart attacks, diabetes, and a stroke. He drives an old four-cylinder pickup that can't go up hills very well. He goes as fast as he can downhill so that he can make it up the other side. Oh yes. He ate chocolate candy and cupcakes all day.

The instructor must have an incredible treasure chest full of stories stored from his 30 years of conducting driving schools and was numb to their implications. He was not as incensed as I was over these few tales. If his students' stories were all documented, I would have a wealth of data points to add to my trend chart, but his notes are discarded after each class. Pity. At the end of the day, his hope for the class was, after reflecting upon the statistics presented and viewing bloody videos, to give us a pause when we chose to violate traffic laws. The interviews alone gave me pause to stay out of California, although any one of these massively distracted and insensitive drivers can be in the car next to me on any highway in any state.

IDENTIFYING THE TREND

These brief snapshots of relatively innocuous events are offered as three small dots on my trend line of declining social accountability. They are insignificant when compared to pandemics and national disasters, but they scream loudly about the symptoms of how we are systematically and inevitably accepting and perfecting mediocrity in our lives and in our business dealings. Without using the terminology or the metaphors of my father and grandfather (that always began with "Back in my day..."), the massive decline in product and service quality is thrown in our faces every day, yet we choose to ignore those assaults against the values that served us well over the first 150 years of our nation's existence.

To wit, my cardiologist is so preoccupied with building a massive practice that he doesn't take the time to deal with patients as individuals. Instead, he commutes from exam room to exam room dealing up the same diagnoses for each faceless name, while maximizing his patient-per-hour ratio. I realize that Marcus Welby was just a TV character, but doctors used to work at a level of competency far beyond the minimum needed to keep them from frequent malpractice suits. We are no longer patients, just victims.

The Big White scenario portrays a dealership and manufacturer that jointly evade any level of accountability for product defects. While most warranty service and product recalls are just inconveniences that we have become accustomed to, our experiences with the SUV, regarding potential personal safety, were dismissed with the same level of concern as a faulty door lock.

Less threatening to life and limb, the dryer service incident screams out loud the state of incompetence that is exhibited by many service companies. It is mind boggling that these businesses can continue to operate, year after year, with inept service people who have to make multiple trips to fix the same problem. It is even more profoundly disturbing to me that the business owner allows mediocre customer service to thrive in their organization with no sensitivity to how badly their company executes its services.

Instead of building a happy customer base, their business model is grounded in positioning a large enough ad in the yellow pages to ensure that new customers stumble on their phone number by clever ad placement. As consumers, we have lowered our expectations enough to allow this appliance repair company to stay in business because the next company in the next banner advertisement in the phone book is likely to be equally poor in its service level. Since the economic bust of 2008, most of these marginal performers are out of the workforce and perhaps among the unemployable.

The driving school experience broadcasts several messages to the world. The first is that these schools exist for the sole purpose of expunging traffic violations from our records, thereby leaving no trace of accountability for our infractions. Even more ironic is that I sent my class certificate to my insurance carrier, who promptly awarded me a lower premium for completing traffic school. I received a pass and a perk for being an inconsiderate driver! What is that all about? Did we evolve that cultural set of values by design or by decay? Either answer to the last question, in my mind, is unacceptable.

The second lesson from driving school is the total lack of accountability exhibited by a random collection of individuals only gathered to evade conviction of a misdemeanor traffic infraction. It amazes me that there is a total lack of contrition over traffic infractions, such as an underage driver chauffeuring a group of other underage children or a psychologist driving on a freeway completely overcome and preoccupied with personal grief. What truly stunned my senses were the background stories that framed the lack of awareness of these people to do potential harm to others. Their verbalized value systems clearly established the scenario whereby obeying traffic laws is now optional. I expected the teenagers' lack of parental guidance, but I was floored by the school bus driver who had evolved a personal value system that justified speeding in a private vehicle but not while driving a school bus. If she could travel forward in time, she could potentially have caused a wreck by crashing her personal vehicle into her school bus.

It was my intent in this chapter to place some itching powder into your underwear to help you get a feel of my concerns for our current state of decayed values. In future chapters, I will take a stab at how we got ourselves into the current state of affairs. After that, I will get into some heavy-duty observations about business, personal, educational, and societal issues that are either the cause or effect of our decline as world leaders.

I will also steer clear of political issues as much as possible. The bookshelves are currently lined with volumes being published daily on what is wrong with our representative republic. Neither will I venture into pop psychology. My conclusions are based on observations of organizational and personal behavior from the perspective of a business advisor whose only motivation is to help us all realize our greatest potential without creating more fanatical cults.

CHAPTER 4
HEADLINES OF SHAME

The anecdotal stories in Chapter 3 are merely a setup to what has happened to business ethics in our Country. If any or all of them incensed you or gave you pause, the following will really infuriate you, especially knowing that it is a brief snapshot of how some companies totally cross over to the dark side of business and approach product safety as a line item on a balance sheet.

My last book, *Foreseeable Risk*, is a desk reference for business leaders like you, who are astute enough to embrace the tenets of avoiding liability and doing no harm. It is a compilation of my experiences as a quality control engineer, business owner, and, for the last decade, an expert witness in products liability and organizational negligence. It is also a guide for those who have been stricken with lawsuits that have crippled their organizations and turned their personal lives into living nightmares.

This chapter is dedicated to providing you the impetus to continue reading this book and to take to heart the imperatives of the Apollo Business Model. Whether you are just challenged by the complexity of business dynamics or looking to turn around a foundering organization, we have, in the pages ahead, remarkable solutions to overwhelming problems. We also hope you will join us in the mission to end the trend toward mediocrity, which has become so pervasive in the first decade of the 21st Century.

The following headlines and summaries are from actual lawsuits in which I have given testimony. The names are withheld because of confidentiality agreements and to protect the guilty.

FAMILY PERISHES IN HOUSE FIRE

Investigators found that an electrical receptacle for a window air conditioner was the origin of a fire that led to the demise of a family as they slept. Forensic science concluded that the outlet was defective and unsafe for its intended use. Our investigation, of the manufacturer, discovered that the outlet was assembled offshore in a facility controlled by a US holding company. The compelling evidence of negligence of the discovery led the judge allowing the plaintiff's representatives to inspect the offshore factory.

There they discovered fabrication, assemblies, and inspection processes that were grossly substandard. The production equipment had been transported from the manufacturing facility of the previous owner in the US and was being operated by workers who had no training and had no idea what they were building.

While the product was UL listed for safety and under ongoing prescribed surveillance in the US, the requirements of the UL listing had been grossly ignored and under inadequate inspection at the offshore facility. Changes made to the product were not discoverable without forensic testing and no testing was done to the product after assembly.

The survivors received cash in a settlement that has been sealed so that the public will never know the dangers of this product. Yes, a version of this product is still for sale all over the US.

ANOTHER FAMILY PERISHES IN HOUSE FIRE

Investigators have traced the cause of a house fire that killed several people to a portable space heater. Forensic investigation confirmed that the electrical defects in the heater were, in fact, the cause of the ignition for the blaze that consumed the residents in their sleep.

The space heater had the brand name of an American company. Attempting to trace the location the heater was manufactured, it was obfuscated by a quagmire of holding companies, importers, and multiple factories with similar products.

In their advertising, the brand company made many claims of extreme detail to quality and safety. With their manufacturer located in Asia, we discovered that the requirements of the UL safety listing for that product had been compromised by using inferior plastics that were unable to contain sparks, which led to combustion of the heater and, ultimately, the surrounding furnishings and structures.

The survivors received cash in a settlement that has been sealed so that the public will never be aware of how dangerous the product is. Yes, a version of this product is still for sale all over the US.

MAN DIES IN HOUSE FIRE

Officials discovered an elderly gentleman dead in his bedroom after a house fire destroyed most of his home. The origin of the fire was traced to an oxygen concentrator used by patients that require supplemental oxygen.

Forensic evidence pointed to the pure oxygen in the machine as the source that allowed the uncontrolled acceleration of the fire. Detailed analysis uncovered that the deceased had been smoking a cigarette and came in contact with the plastic hose carrying oxygen across the room.

While it might be opined that smoking a cigarette in bed while using oxygen was certainly a bad idea, further investigation of the machine's manufacturer disclosed that this was not the first time they had encountered this "issue." In fact, they had been involved in previous litigations and were made aware that a very inexpensive check valve could stop the flame from propagating back to the machine that was generating oxygen.

While the victim may have acted inappropriately, the manufacturer was aware of a way to prevent such misuse but failed to exercise an appropriate standard of care.

The survivors received cash in a settlement that has been sealed so that the public will never be aware of how dangerous the product is. Yes, a version of this product is still for sale all over the US.

YET ANOTHER FAMILY PERISHES IN HOUSE FIRE

Investigators have traced the cause of a house fire that killed several people to a box fan. Forensic investigation confirmed that electrical defects in the fan were, in fact, the cause of the ignition for the blaze that consumed the residents in their sleep.

Our investigation revealed that the defect was the result of a poorly manufactured motor and switch assembly caused by a spark that led to the ignition of the fire. The fan was constructed in the US by a company that was in the process of offshoring the manufacturing and closing their US plant.

The compelling evidence of negligence of this discovery led the judge to allow the plaintiff's representatives to conduct an inspection of the factory. The audit produced evidence that the motor and switch assembly had an alarmingly high failure rate as they were received from the Asian manufacturer. Examination of the purchasing documents showed that the manufacturer anticipated a high failure rate and was given a "failure allowance" in the agreement with the supplier. Defective assemblies were discarded without testing what caused the failures.

We also observed an extraordinary number of fans that were removed from the production line when they failed final inspection and testing. The production line was routinely shut down each afternoon to rework defective fans. There was no awareness that creating such a high number of defective products and then reworking the defects, was a formula for disaster.

The survivors received cash in a settlement that has been sealed so that the public will never be aware of the dangers of the product. It is unknown if the offshore version of this fan is for sale at US retailers.

Even to the untrained eye, a pattern of manufacturers exhibiting an unacceptable standard of care for placing defective electrical products into retail outlets in the US is alarming, if not shocking (pun intended). But wait, we are not through with citing case examples; they are just from my personal experiences with less than a decade of expert witness work.

YOUNGSTER ELECTROCUTED WHILE USING A BATTERY CHARGER

A young man was found unresponsive by first responders in the backyard of his family's home. The youth was barefoot and apparently attempting to connect a battery charger to his go-kart when he was electrocuted. It is theorized that the first jolt knocked the boy to the ground and the metal handle of the device fell across his chest causing the fatal electrical current flow.

Forensic investigations discovered that a poorly positioned wire bundle had come in contact with the cooling fan mechanism inside the charger. Over time, the insulation became frayed and, eventually, the lethal line voltage was conducting to the frame of the device.

We were able to prove that the manufacturer did not exercise an appropriate standard of care in designing and manufacturing the device.

The survivors received cash in a settlement that has been sealed so that the public will never be aware of how dangerous the product is. A version of this product is still for sale all over the US. It is now made of plastic instead of metal.

EVEN YET ANOTHER FAMILY PERISHES IN HOUSE FIRE

Investigators traced the origin of a house fire to a "boom box" in the living room of a home where members of a family perished from smoke inhalation. The CD player and radio were part of an entertainment center that was turned off during the night of the incident. Forensic investigations found that a plastic case from the device had ignited and dripped flaming plastic onto the surrounding furnishings and carpet.

Most entertainment devices are actually "on" even when they are turned off. This is to allow remote control devices to be used to activate the main circuitry. Our investigation determined that the power supply, which was on continually, had defects in design and manufacture, which overheated the electrical connections to the point that they could cause ignition. Also, the plastic housing, used by the Asian manufacturer, was not flame retardant as required by their UL listing.

The survivors received cash in a settlement that has been sealed so that the public will never be aware of how dangerous the product is. Yes, a version of this product is still for sale all over the US.

OIL-FIELD WORKER DIES IN WELL-HEAD ACCIDENT

Company personnel discovered the charred remains of a maintenance technician beside a flaming well head. The worker was performing routine maintenance duties on an old production site when the incident took place.

Forensic investigation traced the source of the ignition to a faulty valve that had been installed on a pipe where the well came out from under the ground. Our investigation discovered that the valve used was not specified for use by the owners of the well. In fact, the chain of custody showed that a distributor had substituted a valve meant only for water wells had been substituted for one that had been certified for high-pressure oil and gas wells. As it moved through the supply chain to the customer, no one handling it discovered the substitution.

Even though the threads of the valve were significantly different from the mating pipe, a previous maintenance technician had somehow managed to cross-thread the valve in place creating the opportunity for the valve to blow out violently, which contributed to the fire ignition.

The survivors received cash in a settlement that has been sealed so that the public will never be aware of how dangerous the product is.

CHILD DIES FROM ADMINISTRATION OF INCORRECT DRUG

Hospital personnel were unable to revive a young girl who was administered an incorrect intravenous medication. The youngster was admitted with breathing problems and an IV was ordered with a medicine that was designed to relieve her congestion. Forensic investigation revealed that the hospital pharmacy had dispensed a medication that was similar in name, but formulated to induce comas in seriously injured adult patients.

Our investigation determined that the automated prescription dispensing machine did not have appropriate safeguards to prevent erroneously dispension of drugs with similar names. Human error contributed to the incorrect product being administered to the patient, but the machine manufacturer shared the liability for the fatality.

The survivors received cash in a settlement that has been sealed so that the public will never be aware of how dangerous the product is.

At this point, I hope you are distressed from reading these stories as I am dealing with them over a protracted legal battle. I have seen too many photos of dead children that were victims of manufacturing and service companies intentionally or unintentionally creating a foreseeable risk for the users of the product. I will conclude this horror tale with a few stories that did not cause fatalities, but did result in serious injury and/or financial harm.

AIRCRAFT COMPONENT MANUFACTURER RECALLS JET ENGINE PARTS

A major supplier of metal components to jet engines announced the recall of certain parts that may have contained contaminates that could cause failure in use. A customer located an included defect that could cause the component to crack and self-destruct inside jet turbine engines.

Forensic investigation showed that there was a minor amount of included contamination that had made its way through all of the inspections and tests at the factory and the components were delivered to a number of customers. Fortunately, none of the defective parts were installed before the defect was discovered.

Our investigation revealed that the manufacturer had been purchasing minor amounts of raw material from a supplier that was not qualified to supply aircraft-grade material. In fact, the manufacturer had selected the supplier more than a decade earlier and had never actually performed their own qualification audits and tests to ensure that the materials received were suitable for use in aircraft parts. The manufacturer's extensive quality control procedures had failed to detect that the supplier was never qualified to supply the materials or held to the ever-increasing level of scrutiny that had been placed on all other suppliers over the years.

In trial, the plaintiff lost their claim that the raw material supplier was responsible for the defects because they did not follow their own procedures for purchasing and testing the raw materials.

WOMAN SERIOUSLY INJURED IN BICYCLE ACCIDENT

A woman sustained serious back and leg injuries when she was thrown off her custom touring bicycle and hit the pavement. The handle bars apparently snapped from their mounting during a routine bike ride.

Forensic investigation determined that a weld had failed on the coupling, which holds the handle-bars, and caused them to suddenly detach, leading to the accident. The bike had been custom built from components from various suppliers selected by the manufacturer.

Our investigations determined that the defective component was manufactured in Asia with specifications provided. Unfortunately, the drawings did not specify how the weld was to be done. In the US, The American Society for Testing Materials (ASTM) provides standards for welding integrity that are imposed on welders to ensure reliability and safety. While such standards exist in other countries, there were no such requirements on the fabrication drawing. Obviously, the welding was substandard.

The woman received cash in a settlement that has been sealed so that the public will never know the dangers of the product. Yes, a version of this product is still available all over the US.

WOMAN SUSTAINS INJURIES AFTER BEING EJECTED FROM A MOVING GOLF CART

First responders treated a woman with lower back injures on a local golf course. Witnesses said that she was ejected from the cart as it was being driven by another woman from one hole to the next.

Forensic investigation revealed that there were no safety belts, handholds, or other restraints present in the cart. Our investigation revealed that there are typically no safety devices required for these off-road-vehicles.

The woman is petite in stature and could not reach the canopy or front end of the cart to hold onto over bumpy and steep terrains. There were no handholds on the seat and the side restraints were too short to grasp.

In discovery, it was revealed that these types of accidents are not uncommon and the manufacturer regularly defended themselves in lawsuits over similar issues. In deposition, a representative of the manufacturer testified that golfers were only interested in easy ingress and egress from the carts, which is why the side restraints were much shorter and smaller than one would find on an office chair. Furthermore, he testified that golfers would never use seat belts and that they would not subject themselves to safety training for operating the carts.

The woman received cash in a settlement that has been sealed so that the public will never be aware of how dangerous the product is. Yes, a version of this product is still used all over the US. Next time you get in a golf cart, determine if you are adequately protected from ejection!

I have many more examples and anecdotes, but I think I have made the point that there are many hidden dangers in manufactured products and services. While our manufacturers, importers, distributors, and retailers may dismiss these types of problems as "acceptable failure rates[4]," I believe (and it is my experience) that we can build products that are intrinsically safe, do no harm, perform to specification, and be sold at a competitive price.

I am calling manufacturers to task and examine their predisposition for greed, make target sales numbers, and look for viable alternatives to producing products that are substandard and dangerous. OFFSHORE MANUFACTURING IS A SYMPOTOM, NOT A SOLUTION.

The imperatives and business models revealed in this book are a wake-

---

[4] In my expert reports I refer to this notion as "acceptable kill rate."

up call for entrepreneurs and leaders, like you, to reexamine the calamities and catastrophes, such as the collapse of Enron, securities giants, banks, and other iconic organizations, and to look beyond conventional wisdom and the acceptable kill rate for a new model of moral and ethical business behavior that creates profit, growth, and sustainability based on providing the best and safest products and services at a competitive price within our own borders.

The 21st Century is unfolding as the era in which we've become a third-world political power and a nation that is incapable of providing its own resources, services, and products when we truly have the potential to once again lead the world in all aspects of commerce.

We made up our minds to go to the moon and accomplished that impossible task in seven years with a coalition of businesses from the size of IBM to two-person machine shops. It is waiting for you to learn and adapt your business and culture before it too late to save your company and, inevitably, the Republic.

CHAPTER 5
HISTORY OF THE TWENTIETH CENTURY

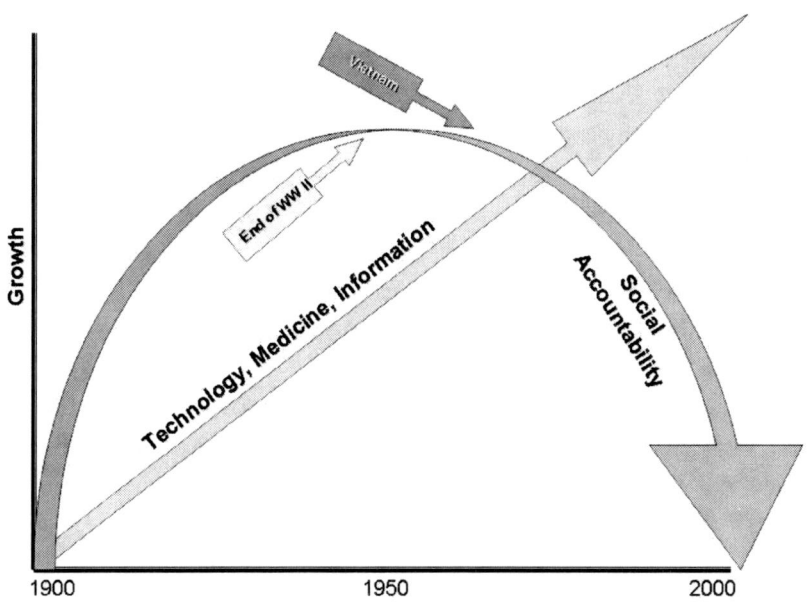

The Bell Curve of the Twentieth Century

The rise and fall of the Roman Empire. The rise and fall of the Third Reich. The rise and fall of Communism. Authors would have us believe that history has defined cycles. If it rises, it must fall, or something to that effect. For statisticians, maybe history is more like a series of curves similar to the societal accountability arc I've depicted above. I'll embrace the latter because any activity that can be categorized as a human "event" often peaks in the middle of its life cycle and then decays (The rise and fall of…). One of the arguments I will put forth in this book is that human behavioral history definitely appears to follow a series of rising and falling curves and we can learn from the past to either turn these trends around before they irreversibly take us to Armageddon or make these curves less steep and less painful.

The second vector on the above chart is the growth of technology. History shows that we continue to advance in science regardless of our societal highs and lows. The evolution from rock knives to automatic weapons gave us the capacity to inflict more damage to mankind during the destructive periods in our historical cycles. If we do not learn from history, the journey from Typhoid Mary to some dissident gaining access to a pandemic virus that can be transmitted across the globe in mere hours makes my impassioned case of learning from history more urgent, if not critical, for the survival of the human race.

As I said earlier, I am not a historian so I will not spend our time together chronicling each step in our growth from an agricultural society to eventually leading the industrialized world. I will, however, recount key events because they set the context by which we've established our insatiable appetite for growth, wealth, leadership, and the inability of humans to maintain social accountability in times of success, particularly in business.

Humans have been successful at building one-off commodities since emerging from caves. We could have built a lot more pyramids with pneumatic jackhammers and overhead cranes, but evolutionary scientific discovery has its own time table. As southern humorist "Brother" Dave Garner used to say, "The cavemen didn't sit around the fire saying: we ain't ever going to have no television at this rate."

If we subscribe to the axiom that "necessity" is the mother of invention, then we must assume that Europe, in the 18th century, was the incubator for the "more and faster" human cravings that created the "necessity" to invent machines of mass production and superhuman strength. Sailing ships and beasts of burden just did not get us where we wanted to be quickly enough, nor did they allow us to extract enough minerals from our mines or crops from our fields. Watt's steam engine (1775) is credited with beginning the move from hay-burning and natural energy sources to mechanical power. With the application of steam engines in tractors, locomotives, and in Fulton's boat, our life in the fast lane was launched.

Over recorded history, each group of our predecessor craftsmen and artisans enlisted apprentices and passed trade skills from generation to generation, ensuring continuity of critical know-how. This tradition ended when we discovered that technology could create much larger quantities of buggy whips at greatly reduced costs by low skilled "machine operators" instead of craftsmen. As society became more urbanized, the need for pyramids and hand-tooled buggy whips passed and the inefficiency of individual craftsmen became an overwhelming impediment to our growing appetite for goods and services.

With credit to the inventors in Europe, The Industrial Revolution might have taken centuries to evolve if American ingenuity had not been added to the mix early on. For instance, we didn't invent muskets, but Eli Whitney (of cotton gin fame) invented interchangeable parts for them. Howe invented the sewing machine, but Singer made it a marketing success. Ford didn't invent the automobile, but he made them widely available. But I am getting ahead of myself.

Americans of this era were direct descendants of malcontents and rebels who had fled European monarchies in search of a better life under self-rule. Immigrants arriving on our shores at the end of the nineteenth century were also fleeing persecution and looking for opportunities in the promised-land. Serendipitously, this confluence of cultures and emerging technology mutated a new phenomenon we now call entrepreneurship. Pre-Industrial Revolution, humans began with stone knives and evolved them into finely crafted sabers. Europeans invented the tools and processes to make these knives and sabers of the highest quality that aristocrats could afford. Americans took the highly-evolved cutlery and asked, "How can I make a lot of these, sell them to people who need it, and make money for myself?"

While Europe was perfecting gas illumination, Americans were entrepreneurially stimulated by Edison's early light bulb (built on the foundation of the inefficient European carbon-arc lamp). Rather than finely crafting the bulb, however, Edison stayed with the program until he invented tungsten filaments. We developed mass-production so that incandescent bulbs could replace gas lamps in every home. Realizing another entrepreneurial need, Edison also launched the first electric generating plant in 1882, just in time to power the light bulbs and, possibly, in anticipation of Tesla's induction electric motor invented in 1888. I wonder if Tesla ever envisioned his motor being used to power an automobile that bore his name! I also wonder what Edison might think if he learned that the manufacture of incandescent bulbs would be outlawed in America in the 21st Century!

The emergence of the mechanical loom and jenny that caused great controversy in the textile industry of Europe created an American entrepreneurial opportunity that has since become the garment district of New York City. Bessmer's genius in steel making was only an incremental evolution in construction until Fuller and Burnham went to Chicago and designed the first skyscraper.

I don't know the answer to the chicken and egg conundrum, so I won't guess at how Scots-born Andrew Carnegie immigrated to Pittsburgh and worked his way from a bobbin-boy in a textile mill to one of our first American entrepreneurial geniuses. I do, however, credit him with creating a paradigm shift as he departed from the single-craft mentality of his forefathers and worked as a boiler tender, a clerk, a messenger, a railroad telegrapher, a labor organizer, the inventor of the railroad sleeping car, a worker in the Union Army, and one of the pioneers of the Pennsylvania oil boom. Eventually, all of these diverse skills would avail him the opportunity to successfully build bridges, telegraph companies, and steel mills, which lead to the amassing of an "obscene" fortune. In Carnegie's success story, I can see the seminal transition from "necessity, being the mother of invention" to parlaying the inventions of others into creating new markets. Just as Ford did with the automobile, entrepreneurial Americans, like Carnegie, summarily transformed the traditional business culture from perfecting craftsmanship for the elite to "If you build it. They will come." and "They will buy."

Another significant series of events associated with Carnegie were his endowment for research and development foundations and funding for international peace initiatives in the Philippines and Central America. Instead of perpetuating the model of his ancestors by filling castles with priceless art treasures, he channeled some of his wealth into creating opportunities for others to enjoy freedom and unlimited entrepreneurial potential. This "spin-off" of Industrial Revolution capitalism would later (in my opinion) become a key element in America's growth to world domination.

As the 20th century opened, Morse's 1836 invention of the telegraph had evolved into the first transatlantic radio communication spawning the information age. The first airplane flew at Kitty Hawk in 1903, which took us to the first mile of a global transportation revolution. While Rudolf Diesel was perfecting his engine in Germany, Ford was inventing imperfect production lines in Detroit and giving us a glimpse of factories in the future. Curiously, in 1905, Einstein was publishing his Theory of

Relativity, while Freud was publishing his Theory of Sexuality. Surely, some societal scholar has connected the dots on these two revelations. Between 1907 and 1909, the first electric washing machine was invented, while the NAACP was also founded. I doubt there is any synergy between those two events.

By 1914, The Great War had commenced in Europe. A year later, 128 Americans died when the Lusitania was sunk by a German U-Boat off the coast of Ireland causing us to intervene in the otherwise-European war. I cite this event as a significant milestone in our evolution because, in less than 150 years, we moved from a rebel colony to an industrial society that was provoked to be involved in a war started over German imperialism within Europe. In my opinion, the same social conscience that Carnegie exhibited when he created his foundations was a national phenomenon that would precipitate our involvement in many more international conflicts over imperialistic inequities and terrorist rulers.

Perhaps another curious coincidence happened in 1919 when World War 1 ended and the 18th Amendment made alcoholic beverages illegal in America. Our soldiers returned home to find an unplanned result of the Industrial Revolution - poverty, crime, and the removal of legal booze was presented as a "noble experiment" for remediation of these social ills. The 1920's Volstead Act exacerbated the crime wave it was supposed to abate. Bootleggers learned the other vices of their customers and organized and expanded their entrepreneurial crime businesses. There may have been a minor "curve" from 1910 to 1930, starting with Carnegie's social conscience on the upside, swinging down to the Mafia-patterned mores prior to the Saint Valentine's Day Massacre, and the stock market crash of 1929.

The 1930's, though scandalized by organized crime, did not mark an end to the growth of social conscience. Proud Americans standing in bread lines, living in crowded tenements, and surviving the dust bowl may have woven a thread into our fabric of nationalistic survival that will triumph forever as a global benchmark. Whether or not you support the politics of FDR, programs like the CCC, WPA, and TVA helped stimulate our national pride and purpose. It is widely believed that World War II actually ended the depression, not the New Deal(s), but our complex history of technological growth in spite of social upheaval is undeniable.

As Europe was being subjugated by imperialism for the second time in the century, America watched closely but kept distance from the active fight. On the other side of the Pacific, we were immersed in dealing with Japan's imperialistic government, which was looking for ways to conquer countries producing the natural resources needed for that island nation to industrialize. All the "observing" came to an end in December, 1941, when Japan invaded the United States and Germany declared war on us, both in the same week. As abhorrent as those events were, they may have been the most significant stimulation for technological growth and social conscience in our history. The decade of the 1940's produced nylon, atomic bombs, rockets, computers, microwave ovens, Polaroid cameras, and jet aircraft, not to mention ball-point-pens. It also united us, perhaps more resolutely than the Great Depression, in a single purpose. Our young men joined the armed services in record numbers while housewives collected paper, sold war bonds, and built airplanes. From what I've read about the first half of the 1940's, we were so preoccupied and unified over defeating the Italians, Nazis, and the Japanese that our melting pot of ethnicity imploded cohesively into one American culture with an incredible resolve and unity. Nationalism and our decisive victories against tyranny and imperialism in Europe and Asia raised our moral and ethical fabric to its pinnacle at mid-century. Some historical scholars believe that we should have stopped the spread of Communism in the 1940's, but we will never know, for sure, the wisdom of ending the War with the fall of the Nazi regime.

In the 1950's we converted rockets from carrying bombs to lifting satellites into orbit. We used the nuclear bomb technology to create nuclear submarines, aircraft carriers, and power-generating plants. We transitioned fighter jets into passenger jets. The expanded GI Bill provided money for servicemen to continue their education and to buy a home; Levittown became the archetype for the American Dream.

Despite our post-war prosperity and nationalism, The Cold War and racial unrest were beginning to divide us as a nation. However, we learned to get along together as Irish, Italian, British, and whatever other European cultures immigrated to Ellis Island. We went brain-dead when it came to getting over the differences between black and white, yellow, red and brown. As a nation, we were only a few short years from the end of slavery, but there were still a lot of Civil War iconoclasts who didn't care about Lincoln freeing the slaves.

We also entered the Korean Conflict. It was the first time we were engaged in a military action that had no clear definition, except to halt the spread of Communist aggression on a small peninsula in Asia. It wasn't a traditional war as we had known in the past. We were familiar with working with our "Allies" in WWII, but in this war, between two neighboring countries, we were ostensibly part of a peacekeeping force. That definition was hard to understand, since MacArthur's troops were on the ground and in the air carrying out traditional military activities against North Korea. After three years, hostilities ended in an "armistice" rather than in a peace treaty or surrender. For the first time the outcome of a conflict was unclear and indecisive for the Americans and the Korean DMZ still exists, 60 years later.

Meanwhile, color TV was introduced, Velcro was invented, and the polio vaccine was created, marching our societal growth forward even more rapidly. The technology machine we created to support WWII and Korea was being retooled for peaceful and "labor saving" products.

I mark this era as the beginning of the downward trend in social accountability. Our freedom from the Depression and two World Wars precipitated a measurable change in our values from constant strife to having a family life. We now had 40-hour work weeks, split-level homes in suburbia, weekend barbeques, and nightly TV shows. Automobiles were again in mass production, affordable, and the Interstate Highway System called us to the Open Road. As Dinah Shore would tell us every week, "See the USA in your Chevrolet." Even with the racial unrest and the occasional air-raid drill, we had become virtually carefree and unencumbered by social discord. As rock and roll music made the scene, our parents admonished us for listening to such heathen lyrics and for watching the wild gyrations of Elvis on the Ed Sullivan Show. They predicted the downfall of Christian and Hebrew morality, as we idolized James Dean instead of Audie Murphy. They admonished us not to repeat the stock market crash of 1929 by our excessive use of credit to buy toys and "labor saving devices."

As the 1960's arrived, a war-hero president was replaced by a suave aristocrat and the White House became Camelot. While the Bay of Pigs reminded us of the ever-present Soviet threat, we were also enamored with JFK's exploits with Marilyn and the Rat Pack and with Jackie's hats and parties. While the Berlin Wall was being erected, we were more intrigued with launching men into space, Cassius Clay and GI Joe toys.

The assassinations of John, Bobby Kennedy, and Martin Luther King, Jr., combined with the growing racial strife, clashes in Germany, Soviet aggression, and the unrest in Indochina, it finally put significant dampers on our 20-year-long celebration of VJ and VE days. We are now better educated, overweight, and more likely to have radical views on political and social issues of the day. Although we've been unified while recovering from the Great Depression and the two World Wars, we became polarized over the results of our affluence and our need for individual aggrandizement. One faction was exploiting capitalism as a means to personal wealth and "personal freedom," while another movement had started to spread peace and love to cure the inequities of the world, with the goal of "freedom for all." While so-called peace-nicks espoused social conscience, their lifestyle was largely invested in drugs, sex, and rock & roll. Their protagonists, the Yuppies, were openly critical of the apparent moral decay modeled by the free spirits and love children, yet these model citizens were busy re-inventing white-collar crimes.

In 1961, The Soviet Union's encouragement of Communist aggression finally caused America to take a position with our "friends" in South Vietnam. We sent "advisors" to Indochina and became enmeshed in a guerrilla war that was not documented in the books that Patton had read. The World Wars were fought under "rules of engagement" honored by even the most ruthless participants (for the most part). The imperialistic assault on South Vietnam had no "rules"; I'm not sure American advisors were really skilled to dispense advice in jungle warfare. JFK's naivety in declaring America the "great defender of freedom," supported by Defense Secretary McNamara's optimism caused us to send more and more "advisors" to Vietnam. By the time LBJ inherited the "conflict" in 1963, the Diem regime had fallen and we were, for the first time, on the losing side of a rebel war where women, children, and morality were all expendable.

There was little investment by our citizenry in this conflict. The middle class didn't see a threat to their way of life. The peace pacifiers were protesting to get us to withdraw. The racial issues at home were adding even more unrest to the equation. By 1964, however, President Johnson had committed to "stay as long as it takes" and we were completely enmeshed in an air, sea, and ground war. The draftees, who were sent to fight, not only lacked the motivation of WWII soldiers, but were delivered to a battlefield where all rules had been suspended either by apathy or by the inhuman tactics modeled by the Viet Cong and NVA. With the absence of rules and the continual exposure to human atrocities, drugs, sex, and rock and roll became the de-facto culture for large segments of our armed forces. The imbedded press kept the troops informed of the lack of support they had back home. Within the USA, the press provided us with graphic coverage of how our troops were being slaughtered by an unseen enemy. The only apparent victories we could gain were under the guise of special operations, covert actions, and guerrilla tactics adopted by our newly morphed "elite forces."

Chinese troops began congregating along the North Vietnamese border with word of inhumane treatment of our POW's circulating, successful coups were being conducted within South Vietnam and the bad situation kept getting worse. From 1965 through 1968, the jungle war raged, casualties climbed, and the anti-war movement had spread to most western capitals. Drug abusing soldiers returned home to protesters and communities, who shunned them instead of saluting them for their military service. They became unwelcome guests at understaffed VA Hospitals instead of being openly invited to be educated under the GI Bill and move to "Levittown". Between those challenged by racial unrest and unskilled veterans returning to a hostile homeland, the land of plenty created after WWII was becoming a major breeding ground for public welfare programs and a widening chasm between the haves and the have-nots.

By the time Nixon and Kissinger inherited the Vietnam mess in 1969, the left-wing press had put their spin on the My Lai massacre and other abhorrent allied acts committed in turning back the Communist scourge. By mid-1969, over half of the American troops were abusing drugs and drug related casualties were outnumbering battlefield casualties. Later that year, there were almost as many peace protesters in Washington as there were American troops on the ground in Vietnam. In 1970, protesters were killed at Kent State, the war moved into Cambodia (yes, it did), our troops were killing their own leaders with fragmentation grenades, while peace talks continued to be unsuccessful. In 1971, the American death toll exceeded 45,000 and our troops were refusing direct orders to engage the enemy. In 1972, peace talks were on and off and Hanoi-Jane Fonda was welcomed into the enemy's capital city. In January, 1973, the so-called Paris Peace Accord[5] was signed, setting the stage for US troop withdrawal, and the unchallenged takeover of South Vietnam by the North (Attention. Those who want to pull out of Iraq before the job is done, we made that mistake 30 years earlier). In 1974, President Ford proclaimed that he was powerless to stop the continued hostilities in Viet-

---

5   Notice how ending wars morphed from surrenders to armistices and now to peace accords? The next iteration might be "promises of understanding."

nam because of the Congressional Ban on military troops being used in Southeast Asia. In April 1975, Saigon fell and America suffered military, political, and social humiliation. The flag on my chart at the beginning of this chapter marks the end of the Vietnam era as a major milestone in our journey of social decay.

The 1970's were also rich in other shameful events like the resignation of Vice President Agnew and President Nixon. Elvis ended his reign as King, unceremoniously by having a heart attack in the bathroom (which the tabloids attributed to drug abuse).

The 1980's took Sally Ride on a ride into space, but the Challenger explosion put the Space program on a long hold. While Ronald Reagan was making headway in uniting the Country, AIDS became a global epidemic. The rich got richer, especially after the Berlin Wall came down and the global markets, which had never been known to us, opened. We developed an insatiable appetite for excess, extreme, and vicarious living. Ed Sullivan and Chubby Checker were trivialized in their entertainment value by Bungee Jumping and Heavy Metal. We had evolved from the Pinto to the first Hummers. Pong was replaced by the voracious Pac Man, which was then eaten alive by Nintendo. Mind challenging video games such as Tetris were replaced by Super Mario and its endless occupation of time with mindless and useless sedentary activities.

The last two decades of the twentieth century was actually a blur of mind-boggling innovations in computer technology that placed a computer in nearly every home and a cell phone in nearly every available ear. In the mid 1980's, we were able to store kilobytes of information. By 2000, we could routinely store gigabytes on our mid-priced white-box computers. In less than two decades, microcomputer technology revolutionized everything from automobiles to washing machines. Commercial satellites, the Internet, and high-speed data access allowed me to research the background for this book in days instead of months.

The end of the century also brought us the Unabomber, the first World Trade Center attack, OJ Simpson's trial of the century, Monica's semen-stained dress, Clinton's impeachment, and other examples of our moral decay and vulnerability to the value system patterning the emerging Generation X. The Gen X moniker was given to the children of the Baby Boomers by Douglas Copeland in his book of the same name. The best definition of a Gen X person I have ever heard goes something to this effect: "Generation X is made up of cynical, hopeless, frustrated, and unmotivated slackers who wear grunge clothing, listen to alternative music, and still live at home because they cannot get real jobs." I am at a loss to paint a more vivid picture of those who are to become the leaders of the 21st century than what is in that one sentence. I won't try, at this point, because I am still chronicling the history of the 20th century that led to the mutation of Gen X (Don't get ahead of yourself. I am aware that not all 21st century youths are Gen X'ers, just as all Baby Boomers are not a product of drugs, sex, and rock and roll).

The 1990's also brought the term "dot com" into our vocabulary. It created unsustainable wealth for some, but failure for most who entered this world with obscene business strategy. Our business whiz kids actually developed a model for startup businesses that was based on money in motion, spending revenue, and creating nothing of value. Instead of traditional businesses that sold products or services of worth, the "dot com" model was based on obtaining investment capital to spend unaccountably and generate revenue streams tied to clever "ideas" for Internet-based businesses, which seldom were successful. Whatever revenues that were generated were paid to the principals in the form of salaries, bonuses, and perks as they tried to bring their ideas to a point where some investor would buy them out or launch an IPO before anyone discovered there was no substance to the business. During the boom days, I worked with venture capitalists who were convinced that a one-out-of-ten success ratio was a winning business strategy with dot com companies. I worked with at least a dozen entrepreneurs who believed that business plans existed only to raise capital and once they had "the money," they could do whatever they wanted to do. I had more than a few Harvard MBA business brokers advise dot com companies to get rid of me because it was my intention to help them build a viable business infrastructure that had metrics and provided value to the clients. In my opinion, the "dot com" boom to bust is the most blatant example of our national obsession with the absence of business morality by building a national business model on self-serving abuse with other people's money and a lack of anticipation of return-on-investment.

As the new century unfolded, I would be remiss if I did not chronicle the historical events of the post-9-11 business world and the next generation of "dot com" Enron- Andersen clusters of business scandals. Let's not forget about the UN Food for Oil mess and the introduction of "extreme" television.

Amidst the tragedy of 9-11, I saw a potential for our nation to rally and unite, much as we did after the invasion of Pearl Harbor in 1941. I started interviewing business people about how they had to reinvent their businesses because of the lessons we had learned from the attack on our financial nerve center. Within 6 months, I had compiled a book proposal that contained stories and lessons learned from the tragedy. My publisher said that (at 6 months) it was old news and no one wanted to read any more about 9-11. His reaction was indicative of how insensitive we had become as a nation and how our social values had decayed over 60 years since the last invasion on American soil.

While living through the unraveling of scandals at Enron, Andersen, WorldCom, Tyco, and the like, one pervasive question kept ringing through my mind: how, as a society, had we created business environments that could allow those levels of abuse and corruption to continue unchecked for such a long period of time? Did we learn any lessons in corporate morality from "dot com?" Moreover, how could so many bright individuals be recruited as operatives in manifestly corrupt and immoral business behaviors? It is incredulous to me that we have evolved into a society that condones conscious deceit and fraud between management, employees, and shareholders. Then, we found out that these scandals are just models for the UN as they pull massively corrupt behavior on an international scale by colluding with a barbaric dictator in the oil-for-food scandal. Altruistic funds not reaching the needy people of the world isn't anything new, but this scam had no chance of ever being anything other than the padding of politicians' bank accounts. It's as if an actuarial team had design the fraud plan in accordance with a formal "requirements" document of avaricious behavior.

As this history is being compiled, we are experiencing a deterioration of support for the so-called War on Terrorism that mirrors the lack of support we experienced for the Vietnam Conflict. In that war, we were introduced to the barbaric manifestation of imperialism. In the current war, we were introduced to a religious cult that is convinced that if you aren't one of them, you must be exterminated. I see radical Islam to be much more compelling of a threat to our society than the communist menace in Indochina. We will find out, in a decade or two, if we will be allowed to eradicate the threat or if we will repeat the mistakes of the past.

In the next chapter, I will relate how these (and other) historical lessons led to the "rise and fall" of our social accountability over the last century. I will overlay other curves of phenomena that I have discovered to support, cause, or track the curve depicted at the beginning of this chapter. These examples include the tragedy we call public schools, the effect the information age has had on our behavior, and how song lyrics and television commercials give us audio and visual proof of my conclusions about social decay. We will eventually discover how that decay has infected and compromised all five of our senses.

One machine can do the work of fifty ordinary men. No machine can do the work of one extraordinary man.
Elbert Hubbard (19th Century author)

THE TRANSITION FROM ACHIEVEMENT TO APATHY

I believe that those who are open to learn from the past are more likely to achieve breakthrough innovation because they do not waste time repeating their own mistakes or the costly errors of others. General George Patton exemplified this model during World War II. The strategy behind his battlefield success was derived by studying the history of war. By being a student of each significant campaign, victory, and defeat, he did not waste time implementing battle plans that had failed in the past. He placed himself in each combat scenario and weighed the lessons learned amid the human tragedy of war. He used that knowledge to create ever-more effective strategic plans. By studying the past, in the heat of battle, he was also able to draw appropriate tactical tools by responding to unforeseen challenges.

Alas, Patton also believed that war exists to serve the agenda of the politicians, which is reflected in his belief that war would inevitably, and forever, repeat itself.

In one of his poems, he wrote:
"So as through a glass, and darkly
the age long strife I see
where I fought in many guises,
many names, but always me."
"So forever in the future,
Shall I battle as of yore,
Dying to be born a fighter,
But to die again, once more."

Without dwelling on his reincarnation theory, he was probably correct about battling forever. Ironically, Patton might have enjoyed the ending of the 1983 movie *War Games,* where a computer playing "global thermonuclear war" concluded that in war, "the only winning move is NOT TO PLAY." Until society has learned that lesson, it is fortuitous for us to have the occasional Patton around when we need one.

In business and in life, the lessons from battles of yore can be critical to act wisely continually without repeating costly mistakes of the past. I'm not suggesting that the answers to our current social and business dilemmas can be solved (as Patton did) by some epiphany derived from the Peloponnesian Wars. I submit that our current "curve" had its launch during the Industrial Revolution and its decay in the last half of the 20th Century and by studying the last hundred years, we will find the causes. With that approach, let's begin studying the journey from accomplishment to apathy with knowledge provided by the history we chronicled in Chapter 5.

NECESSITY IS THE MOTHER

The motivation for mechanization can be anecdotally tied to the European race to colonize Africa, Asia, and any other regions that could be subjugated to support their rapidly growing economies. European imperialism started in the 16th century, although it really picked up steam around the time of the Industrial Revolution (pardon the Robert Fulton pun). Japan was also beginning its colonization dreams. So, most of the emerging industrialized world was developing technology that would help expand their spheres of influence. In the US, however, the Industrial Revolution was not a byproduct of colonialism. We were still at the effect of Manifest Destiny, but we had plenty of real estate in the Western part of our own Country to tame and made only minor excursions from our continental boundaries to places such as Hawaii, the Panama Canal, and the Philippines. As a Country, we also had to deal with the Mexican War and the inconvenience of being massacred by Native Americans, from whom we were stealing land. But that's another story.

The westward migration, Gold Rush, and European immigration were more than adequate stimulus for us to take a principal role in leading technology and innovation. Since we had no single heritage and evolved as a colony, there were no major historical paradigms to be overcome while looking for inventive solutions to challenging problems of growth.

While the British were innovating to get the spices of India to Piccadilly, we were motivated by the unbridled expansion of cities and the need to move West more quickly and efficiently. Our politicians and citizenry unwittingly evolved a serendipitous balance between the pioneering spirit of individual self-reliance and the left-over Manifest Destiny-drive to replicate freedom wherever and whenever we had the opportunity.

THE FIRST THEOREM OF HUMAN ROCKET SCIENCE

I firmly believe that the concept of "entrepreneurship" is best defined by the actions of our early-twentieth-century-malcontent American pioneering ancestors. Their forefathers had been craftsmen and artisans, but that business model limited output and growth potential. Tom's First Theorem of Human Motivation states that: ***The driving force behind entrepreneurial achievement is impatience.*** Therefore, our new breed of business entrepreneurs was motivated by impatience to replace afternoon high-tea and siestas with the ten-hour work day. They must each have gotten a dose of virtual itching powder sewn in the seat of their newly designed Levi 501 denim overalls. Horace Greeley may have been quoted "Go West Young Man," repeatedly, but the early US pioneers of the Industrial Revolution added the adverb "quickly" to this mandate. The pioneering spirit to achieve personal independence, wealth, and freedom combined with "impatience" will turn out to be a formidable power source that has been fueling us (as commentator Paul Harvey used to say us, "U.S.") for the last century.

THE SECOND THEOREM OF HUMAN ROCKET SCIENCE

We are innately aware of the five senses that most of us are blessed with. These senses give us information that can range from pain and horror to pleasure and even ecstasy. They can cause us to instantly change our behavior or modify our values forever. The disgusting smell of odorized natural gas may signal us to dash away from a pending explosion while the smell of baking biscuits may motivate us to get out of bed on a Sunday morning. The sight of a fatal auto wreck may change the way we forever drive our vehicles while the sight of a golden sunset over an azure lake may cause us to phone an old friend to discuss pleasant memories.

My Second Theorem adds a Sixth Sense that is entirely separate from the other five senses because it is internal, rather than external. I call it The

Rush. The *Rush* is the sublime feeling you get when you hit your first home run. It is the pride you feel when you get your driver's license and take the car out on your own for the first time. It is the tingle you get when someone you care about unexpectedly expresses their love for you. It is also (I presume) the feeling you get when you summit Mount Everest or land on the moon. The *Rush* is a fleeting moment when all other senses are overwhelmed and logic and reason are put on hold. Somehow, during the *Rush*, our cognitive emotions stimulate a blinding euphoria that makes information, being provided to us by our five external senses, totally undetectable. During that flash, the laws of physics and common sense can be suspended. Since there are many types of *Rush* as there are stimuli that cause them, we are often at a loss to explain them or why they affect us the way they do. For that fleeting moment, however, all we know is that an inexplicable feeling made us temporarily invincible and fulfilled.

Thus far, I've painted a pretty rosy picture of The *Rush*. Unfortunately, I also submit that The *Rush* is the stimulus for the most heinous crimes ever committed. There is apparently a measurable *Rush* associated with armed robbery, murder, and rape that drives criminally dysfunctional individuals to repeat these acts just as it drives highly motivated individuals to seek a Nobel Prize. The same *Rush* that created the sport of bungee jumping could be a more benign form of the *Rush* that possesses a pedophile to molest children. In other words, Tom's Second Theorem concludes that the *Rush* can be the driving factor for our most outstanding human achievements and our blackest moments in history. The corollary that will weave a thread through this book is that as we evolve as a society, the need for the *Rush* to reach new heights and new lows increases in the positive and negative directions with equal and symmetrical force[6]. The positive *Rush* may lead us to cure cancer, while the negative *Rush* may cause us to unleash a biological pandemic that will kill us before we can find the cure for cancer. The positive *Rush* of pride we feel when a group of employees make a breakthrough on a critical project will, hopefully, become more pervasive than the one an individual feels for successfully embezzling funds from their employer. On a plane that we can practically influence, the positive *Rush* that pushed us up the curve on Page 59 is being diminished by the "extreme" *Rush* that we seem to crave more and more and more. As we will discuss in Chapter 7, the weekly "reality" TV shows provide us with more and more radical, disgusting, and abhorrent behavior because we reach new thresholds

---

6   See the sine wave theory in Chapter VIII

of acceptance that have to be exceeded to keep ratings up. This unchecked appetite for stimulation of our five senses to lead to more intense *Rushes*, in my opinion, is at the root cause of our social decay. The more technology improves our daily lives, the more decadent we our expectations become. The more information we have at our fingertips, the duller we are to the data and the perverse events have to be to satisfy our need for a *Rush* to stimulate one of our five senses.

In 1938, Orson Wells created a *Rush* for thousands of people with the spoken word and crude sound effects on his radio program War of the Worlds. We are too urbane to fall victim to that level of unsophisticated stimulation today. While Edison, also, created an incredible *Rush* with his first moving pictures, LucasFilms can only hope to recover millions of dollars of special effects photography costs in Star Wars Episode 1 if the audience experiences a bigger *Rush* than they did from the previous multi-million dollar Episode 2. If, present day, we can barely get a *Rush* from the most advanced special effects showing graphic human carnage in 3-D wide-screen and surround sound, what expectations have we created for the future? As we move through the history and steps of perfecting mediocrity, I will continually refer to how we demand more and more stimulation of our five senses and how the sixth sense of the *Rush* has caused the morality curve to elevate, peak, and inevitably decay.

BACK TO THE EVOLVING CENTURY

Returning to our lessons from history, we visit again the life of Andrew Carnegie and how he became an archetype for US entrepreneurship. From the writings about him, I can only speculate that he got a *Rush* from mastering each of the varied jobs he attempted. When he had conquered commercial telegraphy, labor organizing, or railroad car design, the aftermath of each *Rush* must have led him to become impatient and seek new challenges. Fortunately for Pennsylvania, his need for new opportunities led him to creating an industry from the crude oil discovered in the State to creating a steel empire in Pittsburgh.

Unfortunately, his 1889 book, *The Gospel of Wealth*, would probably not have been published today because of his position, the wealthy have an obligation to be the stewards of society (most contemporary corporate executives are primarily stewards of their own bonus plans and the stockholders' greed). When he sold his holdings to JP Morgan in 1901, Carnegie began an 18 year philanthropic binge that ultimately channeled

90% of his wealth into foundations and corporations that would benefit society and the future of research. In my opinion, these actions were a seminal contribution to the upward trend in the curve of social accountability that began at the turn of the 20th century.

Without becoming embroiled in the international controversy of "who invented what," it is indisputable that Morse, Edison, The Wright Brothers, Ford, and their contemporaries made reality from dreams and created the basis for others to build entrepreneurial empires. From Morse, many communication kingdoms were spawned. From Edison, scores of power generating dynasties were born. From the Wrights, entrepreneurs started a host of aircraft companies and airlines. From Ford's ideas, others began production-line manufacturing and competing auto-makers even taught a few lessons to Henry when they began painting cars in colors other than black. These successes began a new nationalism that continued the growth of the curve of societal evolution in a nearly vertical (positive) direction. World War I also "helped" the positive direction of the curve because of the nationalism that major war created. After all, we were fighting a recognized enemy, rather than fighting among ourselves.

The curve begins to develop a bit of an arc when Prohibition, its attendant crime, and the stock market crash changed our post-war euphoria into survival tactics. In retrospect, the fight against organized crime and poverty may have perversely stimulated the melting pots of our society to gain more unity and higher moral fabric because we shared common woes. As we worked side-by-side on WPA and CCC projects, Americans prevailed over the dark days of the Depression and built an even stronger societal conscience that would be tested when we became unwilling participants in World War II.

There was no doubt about the motives and needs to enter both theaters of World War II. Once we went to war, the Movietone Newsreels not only provided us with pictures of our young heroes being victorious on the beachheads of France and in the jungles of the Pacific, but they created a new nationalism fostered by the obvious moral high ground of the Allies prevailing over the Nazis and the yellow scourge. As in World War I, the enemy was clearly the bad guy and we were the great white hope. Whether a political Dove or Hawk, war against an enemy that is an obvious threat to a nation's way of life is an incredibly powerful cause that unites the people on common moral high-ground. Unfortunately, we never again enjoyed that level of clarity and fusion.

THE SECOND HALF OF THE CENTURY

In the late 40's, we retooled the defense plants and began creating the Levittown lifestyle for the returning GI's. The VA mortgage and the baby boom would change the landscape of America forever. The worries of the Great Depression and the two World Wars were now behind us and we were focusing our energies on quality of life, the television set, and Thursday night bowling. Our social accountability standards moved from fighting a common enemy to keeping up with the Jones'. We built barbeque pits and fallout shelters, while collecting Raleigh coupons and dishes from the movie theatres[7]. "I and me" were clearly replacing "we" as the focus of our society.

One of the postulates to Tom's Second Theorem is that there is a *Rush* associated with having more, bigger, better, etc., especially when your parents were fortunate to have a roof over their collective heads. This complements Tom's First Theorem nicely because our need for "more" is usually associated with "impatience" that spawned the credit card boom. I want it, I want it bigger and better, and I want it now. This was the legacy we would hand-off to the 1960's.

About the time JFK and Camelot came on the scene, we had created a massive momentum in the pendulum of human behavior. It moved from survival and war, past its center point, and sped toward self-aggrandizement and potential anarchy. Our behavior was manifested with the transition of our music content from Eddie Fisher's love ballads to Elvis' Jailhouse Rock to The Beatles' You Never Give Me Your Money. Cinema began its transition from the clear heroes and villains of High Noon and The Bridge on the River Kwai to the beginnings of depicting shades-of-gray as being fashionable behavior, such as Bob & Carol & Ted & Alice and Easy Rider. My own awareness of the dramatic shift in movie messages did not manifest itself until the 1972 release of The Getaway, where thieves Steve McQueen and Ali McGraw killed a bunch of people and got away with the stolen loot. I remember that film being a learning moment for me and that the "beat generation" of the 1950's had been replaced with a new era of socially acceptable behavior where there were not always negative consequences for immoral and illegal deeds.

The Baby-Boomers were moving pretty high up on Maslow's Hierarchy of Needs and were taking on "causes" to fill the lack of time void

---

7   Raleigh cigarettes came with coupons that smokers collected to redeem for merchandise out of a catalog. Movie theatres attracted customers by giving out dishware on certain nights.

of required contributory activities created by their family affluence. Racial inequity, the discrimination against women in the workplace, and international poverty became the source to get a *Rush* fix for the "hip" generation. The definition of personal freedom transitioned from responsibly pursuing life, liberty, and happiness to free love, drug-modified consciousness, and abdication of social accountability. Power of the people was replaced by power to the people. Communal living replaced individual initiative. "Let it all hang out." was manifested literally in exposed body parts. For the first time, parents became caretakers and abdicated parental controls on TV and daycare, as their insatiable appetite for "more and better" drove the evolution of the two-career-parent households. Substance, meaning, and value of life were replaced with belongings and activities. Actually, "substance" became what kids smoked to get high and what parents got from the doctor to alternately dull and heighten their senses. "Meaning" transitioned into living for some social cause. "Value of Life" took on metrics associated with the accumulation of adult toys. As we discovered, our thirst for excess was insatiable. Debt-to-income-ratio was ignored in deference to feeling good and self-expression.

THE VIETNAM CONNECTION

Since we were dispensing with laws of social accountability at home, our sojourn to Vietnam became the consciousness of our multi-morality society. While we became involved in Indochina to protect our "interests" and our allies, there was never a threat of the Viet Cong invading Santa Monica. While communal living was being tried in isolated settings, the threat of Communism was an abstraction to most people who had not been on the Russian front in World War II. Those of us who were 1A in the draft classification system were looking for any legal (or not) avenue to keep from becoming unwilling participants in the Vietnam conflict. While our fathers flocked to enlistment centers to help protect our borders from clear and present danger, we flocked to universities to get student deferments, hoping the unholy war would be over by the time we finally graduated and left the Tappa Kega Day frat house.

The war we saw every day on TV was not the same one we saw in John Wayne's movies and Victory at Sea programs from World War II. While we were protesting "rules" at home, we were mortified by the visual depiction of the lack of "rules" exhibited in the ruthless conduct of guerrilla warfare. The fact that the rules of conventional warfare had been suspended in Indochina was abhorrent to us, while we were being encouraged by dissident leaders at home to protest rules we disagreed on under the guise of freedom and equality. What a paradoxical dilemma.

To further complicate the quandary, the political landscape was becoming more and more polarized about the war and whether or not we should send in the resources needed to end the conflict or to save face and sweep our lack of common direction under the rug. History shows that the differences between conservative and liberal politics were not that pronounced in the JFK White House. Presidents Johnson, Nixon, Ford, and Carter changed that degree of polarization as they bounced American policy from hawk to dove and back so many times that our enemies followed their Communist teachings and waited for America to destroy itself by the conflict within. For the first time in our history, we were "0-1-1[8]" and appeared to be on a path to prove the Communists correct in their prediction of what the "enemy within" was doing to us.

Yes, I am aware of my repetition of history lessons, but I am convinced that we do not take them to heart in managing our own life and business journeys. I am going to continue to reinforce them in the ensuing chapters, adding the powerful influence of the *Rush* before we get to the turnaround messages.

THE YUPPIES

By the end of the sixties, the curve of social conscience and accountability was well into its downward spiral. The embarrassing loss in Vietnam created the fertile ground for mediocrity to grow roots. The evolving yuppies (prior flower children) nurtured the garden of mediocrity by becoming obsessed with personal gain and converting Levittown into gated communities, where Gen-X children could grow up creating their own moral compasses influenced by Charles Manson. By the third lunar landing, networks were no longer devoting major coverage to some of mankind's greatest moments of achievement, but there were plenty of airtime devoted to the Watergate break-in and the scandal of a President resigning in disgrace.

Also during the 70's, our social conscience was forever changed with the legalization of abortion and the birth of the first test-tube baby. Whether you are pro or con on the issue of abortion, it's significant to our

---

8 A quote by Clint Eastwood from the movie Heartbreak Ridge, statistically representing that we had not won a war since World War two (0), we tied in Korea (1) and lost in Vietnam (1).

moral fiber, our government was now involved in family decisions and post-conception-contraception, had become a viable option for cleansing ones-self from promiscuity.

Cheech and Chong were glorifying marijuana, which many were hoping was right behind abortion in the queue for legalization. The misguided morality that led to Prohibition had somehow become the faintest of memories, while George Carlin (who was "high" most of his brilliant career) offered the theory that the Ten Commandments should appropriately be renamed The Ten Suggestions.

I am a devout agnostic, but the direction we were heading, as a society, by 1980 was "unholy," even in my belief system. Vietnam was portrayed as a symbol of human carnage committed by self-medicated soldiers who had no leadership or will to win. When they returned home, they were treated as lepers instead of returning heroes. We did not know it at the time, but it wasn't terribly uncommon for priests to be molesting young children. A sitting President had committed felonious activities, was caught, and resigned in disgrace. Jim Jones convinced a group of religious fanatics to drink poisoned Kool-Aid. Whatever moral fabric we had woven in the first 50 years of the century unraveled in our own hands.

As the new decade of the 80's unfolded, there were attempted assassinations of the Pope, President Reagan, and the successful assassination of John Lennon. Shoppers were trampling each other trying to purchase Cabbage Patch Kids, while we found it necessary to assign age and content ratings to films. Although it was difficult for youngsters to get in to see an R movie, VCR's were in every home, spawning a new industry that made pornographic videotapes almost as common as beer. The *Rush* created by the movie *Deep Throat* in 1972 evolved the need for sick producers to either make or fake "snuff" films depicting murder and dismemberment. How had we made the transition from the *Rush* of a first date to the *Rush* of seeing our first decapitation movie?

ENTER THE PERSONAL COMPUTER

Then, the greatest technological advance of our time, the IBM PC, was introduced in 1982 and our world changed forever. Anyone who had ever used a typewriter or mainframe computer immediately saw the unlimited possibilities of human advancement that could be realized by having a personal computer in every home and office. The ability to store and manipulate data at-will would be as revolutionary as the automobile was to the horse and buggy. Composing the written word and being able to edit and store it at another location would immediately save millions of man-hours of manual typing and filing. The ability of spread sheets to allow us to manipulate multiple rows and columns of mathematical calculations would quickly improve productivity from controlling inventory and balancing the family budget.

Our predilection for "more and faster" was only in its infancy before the PC. Those who had the foresight to realize the boundless markets that could be created by rapid data manipulation, data sharing, visual stimulation, and simplification of mundane tasks became the pioneers of monthly technologic breakthroughs. The velocity PC's added to our lives paled all previous inventions. The speed at which we could "push" technology and "pull" the evolution of software applications was unprecedented in any previous human endeavors.

While it had taken a decade to build the technology to fly to the moon, it took only months to increase computer processor speeds, memory, and storage capacity by orders of magnitude. The practical life of a computer became reduced to about 18 months before it was "necessary" to upgrade hardware and software. PC gurus became more common than auto mechanics. While dozens of programs were being released every month, other geniuses were inventing the technology of "connectivity," where multiple computers could communicate with each other locally (LAN – Local Area Network) or over long distances (WAN – Wide Area Networks). While we had primitive connectivity during Project Apollo, the computers in Houston, Washington, and Cape Canaveral talked to each other over dedicated hard-wired phone lines. Such technology would not do if there were to be a PC in every office and den. Thus, we pushed technology until "servers" were robust and international connectivity became practical - thanks to Al Gore's invention of the Internet!

At this point, the evolution of the computer as a spin-off of space technology was seen as an incredible boon to technology, business, communications, and transportation. The use of PC technology in medicine was directly contributing to better health and increased longevity. Our world of entertainment changed forever with computer-generated special effects and computer games. CD's and DVD's put information and entertainment at our fingertips. Microcomputers are now flushing our toilets and dispensing cash to us at the mall. How can I be so upbeat about computer technology and be ready to make a connection with It WAS Rocket Science?

THE THIRD THEOREM OF HUMAN ROCKET SCIENCE

Tom's Third Theorem is that *PC technology has immeasurably improved our lives, while it has also helped prostitute our value system.* Specifically, the massive improvements in software & hardware technology, connectivity, and information availability have been both blessings and demons. Let's examine each of the two areas individually.

The velocity model we created to expedite "more and faster" computers sacrificed reliability and value for expediency. While the development of breakthrough hardware technology is commonplace, the design model moved from building reliability into products to the commodity becoming disposable. In other words, we created an evolutionary model that has little need for longevity, serviceability, and reliability because the next generation of product is only 18 months away[9]. Instead of creating heirloom craftsmanship with architecture that can be upgraded as technology advances, we are creating bone yards of discarded hardware.

Whoa, Hoss. Don't throw this book at me just yet. I am not proposing that our technology paradigm is a bad thing; it's just different than previous technology models. The reason I include disposable technology into the perfecting of mediocrity is that it created a new mindset where disposability became a way of life in many of our value systems. Just as a 1987 IBM PC-AT has become a relic, the handshake became a relic as an ethical commitment between humans. We dispose of relationships as casually as we do with computer monitors. I submit that we have crafted a set of moral standards that evolves, for our convenience, as quickly and easily as we evolved from vinyl records to compact discs. But wait, my allegations become more profound as we complement hardware evolution with software development.

For those of us who are aware of the amazing advancements we have made in the physics of evolving more and more sophisticated hardware, those enhancements pale in comparison to the complexity of developing the complementary software to run these devices. Computer programs are now more than complex compilations of coded series of electrical-

---

9   From my work as a Senior Consultant to Dell Computer, it is my theory that the genius of the meteoric growth of Dell was in part due to their collaboration with Intel to create faster and faster processors.

ly-created ones and zeroes (voltage turned off and on) connected in an order that causes computers to (hopefully) behave in predicable fashions. While hardware is frozen into a stable and hard-wired format when its design is committed to production, the juxtaposition of a one or a zero in the software can be changed on -the -fly as the function of the software continually changes (mutates). While hardware is bound by immutable laws of physics, the behavior of complex software has no physical boundaries. The more complex programs become, the more likely the effect of a misplaced one or zero can go undetected for months. Software "glitches" like this are seldom discovered in the development and test process. They are often not found until a user performs some obscure function causing the program to perform unexpectedly, leading to potential data corruption. Just as I do not believe that hardware engineers sit in design review meetings where they conspire to build crappy hardware because it has a useful life of 18 months, I do not believe that software designers conspire to release software that makes the most patient scream at the screen when a piece of software crashes and we lose a day's work. I do believe that PC hardware and software create new paradigms of what a level of reliable products should have before they are sold to the public. Software is released when it can generate revenue, not when it is through exhaustive testing. An unintended result is that the ethical standards of product performance have reached a new hallmark of mediocrity.

The first time I visited one of the major mainframe computer manufacturers in the 1970's, I was struck in awe by a large room full of computer tapes immediately adjacent to the lobby. I asked my escort what the room was for. He explained that it was their software library. When I asked how it was run, I received the following explanation. Organizations that purchased their computers also purchased an annual subscription for a "software update service." As the company developed software "enhancements," subscribers would receive appropriate revisions to their existing software, along with instructions on how to patch the software into the existing system. As a part of this "paid" service, customers can also contact their call center to report problems with the software and receive instructions for workarounds. With great glee, my host explained that most of the software "enhancements" were, in fact, fixes for bugs reported by the customers. Workarounds for problems that were yet fixed were sent to subscribers as proactive operational "tips" and pending future "enhancements." I think this was about the same time I heard one of my favorite software limericks for the first time: "Don't fix it. Turn it into a feature." My immediate (but unstated) impression was to make a parallel between their software subscription service and the Mafia protection rackets, where business owners paid for protection from problems they did not have, unless they did not pay the mob. I had to give them credit for creating an elegant profit center for fixing their own software bugs and leaving the customer virtually no option but to buy their annual subscription services to keep their computers operating reliably.

The major software companies evolved this scam into a well-choreographed process for creating a perpetual income stream after selling one software program. It works this way. As consumers, we have been convinced that developing bullet-proof software is virtually impossible because of the endless permutations of hardware and environments users will use to host and run the software. In the popular Windows® environment, we have been brainwashed to believe that the Microsoft Operating System is as robust and stable as is humanly possible because of the inconsistency of quality among outside software developers[10]. The fact that they must continually evolve the operating system, while millions of people use it daily, is the inevitable result of our passion for "more and faster," not a manifestation of the incomplete development and testing of Windows XP®, according to the Gates empire. The software developers claim that the problems we encounter using their programs is the result of the instability of the Windows® Operating System. According to software developers, as consumers, we are contributors to the mediocrity of software because of our insatiable appetite for more and more features more quickly. The fact that we continue to spend billions of dollars on software "upgrades" confirms the fact that we have become culturally accustomed to paying dearly for the mediocre performances of our personal computers. I submit that the fact that we accept the "blue screen of death," data crashes, and unstable software has transmuted our value system and corrupted our expectations for all business contracts. While the PC allowed me to write 12 books in a fraction of the time I could have with a

---

10   I recently capitulated and decided that I needed to purchase a tablet for traveling as my laptop was too inconvenient to use on an airplane. First, it took me from Reno to Ft. Lauderdale to figure out the basic operation of it and Windows 8. Second, three months later, I was having massive problems with MS Outlook. My IT guru told me I would have to install Windows Update 8.1 just to get the tablet stable again. I sometimes fantasize about being part of a class action law suit to regain the lost time we each spend getting and keeping our computerized devices and software working.

typewriter, I would conversely like to find a way to recoup the thousands of hours I spent, since 1987, being a computer troubleshooter and repair person when I only wanted to be a computer "user." I am also dismayed that I will forever be an unwilling beta test site for the largest software developers, as the bugs and crashes I regularly endure are now automatically reported to an unseen collection barrel at Microsoft. At the end of the day, we allowed the mistakes of software developers to be turned into features we pay for and have adopted this sub optimization as our new business model of ethical behavior. Amazing how quickly we adapt to our environment.

The final insult to achieving a stable computer system is Windows Updates. These automated assaults on our computers, veiled in the pretext of keeping us up to date and free from attacks, changes settings in obtuse ways that often requires my IT professional to undo. Please, do not send me any more automated "features!"

BECOMING CONNECTED

The second mixed blessing of the PC revolution is connectivity. As a teenager, I recall making contact on my ham radio set with a school teacher in Senegal, Africa. My imagination was operating at full speed as I created a surreal mental picture of the home and operating conditions that he described through the crackling static. Today, I can converse clearly with hams in Senegal using the latest digital signal processing in my radio set. I can also "keyboard" with them, in real time, over the Internet, while looking at color images from each other's streaming video cameras. Over my digitally-based satellite TV, I have seen a number of historical and geographical programs about Senegal. Not much is left to my imagination any more.

Again, don't misunderstand my reflections. I am blessed to be able to talk with hams in every corner of the world, exchange email with them, and watch television programs that take me into the corners of the universe that I will never get to visit. The lack of information I had at age 13 (that challenged me to use my imagination and want to know more) has been replaced with access to an overwhelming amount of information that challenges my imagination even further to sort out the contradictory data available about any given topic. I can instantly access topographical satellite photos of Senegal as well as detailed photos of the flora and fauna. With the same number of keystrokes, I can access pictures of women having sex with animals and view secret webcams set up in public restrooms. Controlling what information I access, I can choose not to surf the web for pornography. I can't, however, imagine what it must be like to be 13-years-old and have access to the endless variety of information available over the Internet. Many in my generation saw their first pictures of bare-breasted women in National Geographic Magazine. Clever pre-pubescent computer users can now access the entire spectrum of Kama Sutra sex positions at kamasutrafree.com. If he or she is not encumbered with parental controls, they can access the deepest philosophical writings of Sartre or the deepest perversions of sadomasochistic sex. Observe how youngsters currently conduct themselves at movies, restaurants, and the mall. Is our society capable of dealing with the unlimited amount of information provided to us through our PC and Satellite TV connectivity? Just as our hardware and software development processes have degraded to fit our need for "more and faster," our culture degraded because we have not evolved any prerequisites for exposure to extreme information leading to some of the abhorrent behavior that stimulated the writing of this book. I am not doing this for social commentary. THESE ARE YOUR CURRENT AND FUTURE EMPLOYEES!

Most of the world has an education system similar to our K-12 structure where we are gradually exposed to more and more complex information as our capacity to absorb information matures. We are also (typically) not given more information than we are capable of acting on responsibly. This maturity-based system isn't perfect, but it has kept us from total anarchy. Today, first graders can access massive amounts of data that are beyond their capacity to comprehend, both on TV and the Internet. I submit that their lack of personal accountability is the result of being exposed to web sites and TV shows that glorify irresponsible behavior. I will make my case for these assertions in later chapters. These are the individuals who can become the irresponsible adults who watch more and more examples of irresponsible behavior and then, transfer those behaviors with their interaction with the real world; these are your potential future employees. Not only is this extreme behavior becoming commonplace, we are perfecting it.

VOLUME 2
PERFECTING MEDIOCRITY

In Chapter 5, I covered the dot.com scam, 9-11, and corporate fraud in enough detail to remind us how the lack of accountability has become a national epidemic. I have created an overview of how, in my opinion, we have reached ebb tide in social accountability and have yet to mention Bill Clinton's impeachment and his relationship with Monica.

I was going to title this section of the book The Direct Deposit Employee. That term came from an enlightened friend. He first made me aware of what happened to our workforce since the inception of direct-deposit paychecks. In essence, since employees never actually are handed the money they earned and it shows up in their bank account every two weeks, any shred of risk and/or reward for doing a good job has vanished. Ponder that while you reflect on why your workers are not highly motivated and why they dodge personal accountability. Please consider this conundrum as you see more examples of our predisposition for perfecting mediocrity.

We will now take a closer look at specific segments of our lives that encourage avoiding accountability in business and personal life. We have just reached the point (late 2013) where the occupation with the largest number of US citizens is now "unemployment." Who is going to show up at your next job fair?

I am hopeful that the examples and lessons I have and present will allow us to harvest the knowledge and wisdom to take action to turn the tides.

> *Don't worry about the world coming to an end today. It's already tomorrow in Australia.*
>
> Charles Schultz

CHAPTER 6
PUBLIC SCHOOLS: THE FOUNDATION OF MEDIOCRITY

We, management consultants, think that we have the answers to most businesses problems within customer service, process, personnel, and leadership. There is one conundrum, however, that we will likely never solve and it stirs controversy every time the subject is raised. In schools (K-12 and undergraduate college), who is the customer? Is it the student who is receiving the services of the schools? Is it the parents who are paying for the education? Is it society that is the ultimate beneficiary of educated citizens?

We are proving, everyday, that the most successful organizations are those that listen to the voice of the customer (Lexus, Starbucks, etc.) and base their services and continual improvement models on the feedback they receive. Those organizations assign metrics to key customer satisfaction indices and prioritize their future activities based on careful balance of customer need, advancing technology, and potential return-on-investment. I recently had a tour of the new Starbucks roasting plant in Minden, NV. Everyone I interviewed had a clear sense of the feedback they received from the retail outlets and their customers. I leased a new 2006 Lexus SUV and the only compliant I had lodged about my 2003 model had been fixed on the 2006 (cup holders that did not hold drinks securely). If we agree that the customer should drive product and service excellence, how can we do it in schools?

If the student is the customer driving process design, much of the school day might revolve around hip-hop music, lunch, and sports. If the parent is the customer, activities might accelerate the current trend in thinking that the **school should be raising the kids** as well as educating them. If society is the customer, each industry and special-interest-group would have an opinion of what the profile of a graduate would look like. Today, school boards make these strategic and tactical decisions based on their collective predispositions, parent pressure, technology, and budget. There are also regional influences that direct schools. I'm sure that agendas of the school boards of East Los Angeles and of Scottsdale, Arizona are quite different in setting priorities for resources and activities. In Texas, where my boys were raised, I'm pretty sure the school district's priorities were football, baseball, and then academics. Graduating young folks, who were ready to assume roles of responsibility in society, was definitely at the bottom of the priority list unless those responsibilities included deer and duck hunting skills.

If you are wondering why schools are worthy of an entire chapter, it is mainly because education has, for the most part, followed the rise and decay curve since the 1900's in synchronization with social accountability. Also, the school system I just mentioned, graduated my youngest son when he was functionally illiterate. My ex was highly involved in school activities and I, unfortunately, abdicated most of them for my career. I was not aware of how poor his academics were until I read a letter he wrote to my mother when he graduated from high school. The misspellings were rampant, the syntax was horrible, and the grammar was worse. He was a jock and made it through academics at the influence of his coaches and his mother. I also found out, years later, that he was held back a year so that he would be physically larger for high-school sports. I have no doubt that academics would have been enough for him repeat a grade, but I am appalled that sports was part of the equation. I was even more disgusted to find out that there was an existing program for jocks to attend junior college for one year to take "college preparatory" coursework. In other words, the first year of college was designed to teach them what they should have learned in high school, which was sufficient to get into a 4-year college. Fortunately, he turned out to be a responsible business-person and parent. He worked his way up from washing cars at Hertz to managing a regional car-parts empire, but his potential will be forever limited by his inadequate education (he never finished the first year of junior college).

In the first half of the twentieth century, families typically sacrificed their living standards to make sure that their children got a sound education. If you made it through high school, you certainly had enough education to contribute to society. Mid-century, it was common for families to move jobs and households to communities that boasted superior school systems. Many hocked family assets to put their children through college. Because of my independent nature, I worked at least 35 hours a week to put myself through college. Near the end of the 20th century, most schools became a babysitting service and a teacher of the social skills most prior generations had learned from their families. Besides academics, I learned to drive and to type. Today, our children are taught the fundamentals of sex, nutrition, social behavior, and hygiene. What's wrong with this picture? While we debate who the customer of schools may be, there is no doubt that the curriculum was adapted to supply training formerly provided by parents who were materially involved in raising their kids. **We used to raise kids to be prepared to enter society; now we raise them to avoid the negative influences of society.** As my grandsons say, "Wassup?" or in English, how did we drive ourselves down this road of societal degradation?

Just so you know, I believe that students, parents, and society are all customers of schooling and we must work together to define the mission and needed outcome of public education. Based on what our evolving society needs to continue flourishing, we must decide what the profile of a high school graduate and a baccalaureate college graduate looks like, what metrics we use to determine acceptable matriculation rates, and how we go about designing curricula to support our collective needs and expectations. **Left in its current state, public education is going to help us perfect mediocrity and expedite us into becoming a third-world society.**

I do not make these allegations frivolously or without first-hand experience, nor do I seek to diminish the good work being done in many public schools. Unfortunately, however, I see the ill-prepared product of our public schools almost every day in my consulting work and in the training sessions I conduct. Illiteracy is rampant, learning skills are underdeveloped, and behavioral skills are antisocial. On the other side of my experiential tool kit, I am reminded daily, by my wife, of what is going on in public education. She is a Special Education Aide for profound special-needs students.

The year she started working at our local high school, I attended "senior night" with her. Her charge, a severely retarded young man, was "graduating" from the school system because he turned 21 years of age.

The program was very touching because he is a loving individual who was grateful for the attention he got at the ceremony. The rest of the evening was disgraceful. The "honors" bestowed on the seniors were for the most inane activities. The best dressed were in jeans with no holes in them. The worst dressed had bare midriffs and jeans hanging low enough to expose their underwear. One "couple" came to the podium to receive an "honor" in an embrace that included having one hand down each other's pants. I asked the principal about their behavior over coffee and he replied that the best they can do these days is to keep the kids from exchanging bodily fluids on campus! My wife's current charge is an eleven-year-old with a "behavioral disorder." He reminds her daily that his job is to drive her crazy and get her to quit, as he has done with some of his previous aides. By the way, in our State, if a parent requests a dedicated teacher's aide for their "challenged" children, the school must provide one and we pay the bill via federal grants. I ask again, is this the learning model that we have intentionally established for our public schools or have we mutated to accommodate parents who are no longer accountable for raising the children they conceived? Perhaps the greater question is: Has society completely abdicated its standards for social accountability?

Yes, I have met highly motivated youngsters in the last decade. Yes, I have seen recent college graduates who have great moral fabric and are relentlessly pursuing excellence in everything they do. Unfortunately, they are the exception, not the mainstream. Although they make me proud, these few could seize the environment of mediocrity and conceivably evolve a society where they become the elite leaders, while the majority degrades to a non-participatory proletariat. Does the Communist Manifesto come to mind in this scenario? Watch a few episodes of The Apprentice and ask yourself: if we are not growing a new elite upper class of selfish and ruthless business executives, who will effectively manipulate our vacillating political leadership to their own advantage? Ask yourself how the entire senior management staffs of Enron and WorldCom could have, in good conscience, engineered the incredibly diabolical plots that led to their demise? Do we ever learn from history? Why did we have to legislatively create the Sarbanes-Oxley act just to require businesses to act morally and be accountably to their workers and shareholders? Okay, enough already, back to school.

If we were given the opportunity to conduct a completely unbiased survey of our country and to design the roles and responsibilities of our public schools, what do you suppose the results might be? Since this is a hypothetical question with an answer that may be unknowable, let me offer some specifics that would likely <u>not</u> be part of the future success specifications.

The first activity, not making my list of "keepers," might be school boards deciding how to educate children. But Tom, we elect the brightest education professionals to be members of our school boards! Maybe we should be asking our business leaders, regional planners, and futurists to give us some input as to what skill sets we will need in the next decade or two. Perhaps we should observe what is going on with the technological revolutions, the massive growth in China and in the Emirates, and plan courseware on how to reverse the trends of our industries moving offshore. Just maybe, we should look at why it is more profitable to operate call centers in India than in Indiana and begin creating curricula for solving this problem. Maybe we should study the trade deficit, the brilliance of China developing an actual "business language" that will make that awaking giant even stronger (Simplified Chinese), the harmonization of cultures being created in the European Union, and the effect of radical Islam on world security as the basis for preparing our youngsters to take over the leadership of our Country. Our current model of running public schools must be turned from the traditional "push" model (We are the educators. We will decide what to teach.) to a "pull" model where schools boards plan their futures around the stated and predicted needs of our society.

Next on my list to delete is the Bill of Rights we have created for youngsters who are too immature to understand them and have not been given the foundation of accountability to use these freedoms wisely. Apparently, the current generation was born with some endowment policy that came out of the birth canal in a plastic bag. I even asked my son to save it for me when my grandson was born. I would like to read it because I hear it quoted so often. Children and adolescents are apparently endowed with inalienable rights such as behaving however their mood strikes them or how their peers expect them to behave. They have the right to question all authority, to not follow any rules of conduct that they deem to impinge on their personal freedoms, and to blame any negative results of their activities on someone else. Oh, yes. They have a schedule of rights tied to their age that require certain possessions to be available on schedule and without any prerequisites. These include video games, smart phones, televisions, computers, the latest toys, the latest clothes, bikes, skateboards,

ATV's, motorcycles, sports equipment, dance lessons, contraceptives, a chauffeur-on-demand before 16, and a car on their 16th birthday. I am beginning to understand that this is the minimum we must provide to our children under the endowment-at-birth program. Oh, yes. I am told this document also contains their bill of rights, which includes a full range of personal rights, freedoms, and protection by the legal system that reverts blame on their parents and teachers when they commit antisocial acts (See the Driving School story in Chapter 1). Lastly, the document contains an "upgrade" stipulation requiring the latest revision of games, music, software, TV channels, and cell-phone services be given to them as soon as they are available. It's kind of an endowed software subscription service! What is really unbelievable is that the collective WE (parents) have enabled this behavior to flourish and become the norm.

The next bastion that should not make the survey of how we run our schools in the future is the teachers' unions. Unions may have served a great purpose when sweat shops and child labor were rampant. They may have had meaning when there were large numbers of workers performing the same tasks, who could not bargain for themselves. Unions, however, only benefited the workers before the union mission changed to becoming political advocacies and powerful lobby groups and created high-paid jobs for union bosses. Without starting a global union debate, I would like to know why those who are tasked with educating our children need representation to express their needs to management. Surely we would expect teachers to be among our best communicators, wouldn't we? Unions in schools have created the seniority system that helps to perfect mediocrity in education. **Tenured teachers and professors are, by design, destined to be the torch bearers of the status quo and of mediocrity!** If public schools provide the appropriate trained workforce to meet our growing national needs, why do we not learn from the most successful businesses that pay-for-performance, reward-for-innovation, and continually push the envelope of excellence as the key to our future greatness? How can we have superior schooling when we create tenured environments that attract those who need protection from their own incompetent performance? **Seniority breeds apathy, not wisdom!** School districts are unable to hire the youngest and the brightest until the oldest and the least visionary retires.

I feel another disclaimer coming. Yes. I do know very bright and very motivated teachers with gray hair. They are, however, becoming an anachronism. Their creativity is stifled by the restrictions of the school boards. Their ability to create the magical learning environment that is within their beautiful souls is stifled by their inability to control behavior among the entitled youths in their charge. They are helpless in obtaining the funds they need for innovation because the school budget is being channeled to sports programs and to keeping kids from bringing weapons to school. Oh yes. The schools' funding is also used to test teachers, who often do not meet minimum standards of training. Define "mediocrity."

The perceived reward for tenure and seniority is a generous pension and lifetime health benefits. Unfortunately, pensions are evaporating before the baby-boomers are retiring. Health care costs have become prohibitive, so the promise of future rewards for a lifetime of teaching is becoming folly. Define "irony."

The final offer on my current list of what should not be on our list of future school success models is what I call SLOB. That is a crude acronym for parents' selective outrage and blindness. Here are some examples:

- How dare you discipline my child? Why is my child being kept after school in detention?
- You must do something to improve my child's grades. You can't keep my child out of soccer because of his grades.
- My child is getting obese because of the food you serve at school. How dare you take the vending machines out of school?
- My daughter lost her virginity because you teach sex education in school and promote promiscuity. We need a special program for unwed pregnant high school students.
- My child has ADD. You must provide special assistance for him. Why is my child kept out of the mainstream?
- I am a single parent. You must have pre-school and post-school activities available for my child. I never see my child.

Parenting has become an elective function in our mediocre society. PTA meetings used to be a forum for interchange of information. I haven't been to a PTA meeting in many years, but I would be stunned if half of the parents who attend the meetings went to help advance the learning process as there are parents who attend school sporting events.

Just as we have created the environment that encourages mediocrity in business, we certainly have enabled and encouraged mediocrity in public schools either by abdication, apathy, inattention to our parenting responsibilities, or all of the above. We are creating a future work environment for those who believe they are entitled to be supported by the government. Wake up America.

The Entitlement Generation

CHAPTER 7
THE MEDIA: OUR MIRRORS OR OUR TEACHERS?

I am not a media critic, but I include this chapter because it is another mirror of the wide variety of belief systems you have working in your organization. If you continually ponder why your staff is not performing to expectations and do not share your values, here is a glimpse into one of the most audible and visible reflections of the deterioration in our social morality, and, hence, our work ethic.

I am not sure that I will be able to empirically answer the title question in this chapter, but I will certainly give us plenty to ponder when I ask: **Is television and radio a reflection of our society or do they influence our behavior by what they broadcast?** Specifically, I am addressing the programs and commercials on broadcast/cable TV, song lyrics, programs, and commercials on AM and FM broadcast radio. I am opening up with questions for us to ponder such as: Does NBC, CBS, and FOX report the news or make the news? Is Puff the Magic Dragon a song about smoking marijuana? Did A Clockwork Orange define depravity for us or did our society define the depraved plot of the movie? Are commercials for class-action lawsuit attorneys (such as "Were you exposed to benzene in your work place? Call Attorney XYZ") the impetus behind so many class action law suits or are they just filling a demand for legal representation? Do the Crime Scene Investigation shows stimulate us to create more forensically challenging crimes or do we just want to know the graphic details of what happens inside our local morgue? Was the Sopranos just another mob movie or is it a reflection of typical suburban life in Northern New Jersey?

A continual problem for we, researchers, is to step outside our paradigms and assess the objective data we are uncovering, rather than "seeing" the data that only supports our expectations. Having performed scores of "root-cause-analyses" as a Quality Engineer, I have uncovered many problems that had nothing to do with what my predisposition would have predicted, so I am usually not surprised that the facts lead to some unforeseen revelation. In the case of Radio & TV, I was pretty certain that there was a measurable degradation in what we are exposed to in music, programming, and commercials over the last 50 years. I did not expect the data set to show that, by any norms of social accountability, the downward arc has reached an exponential velocity in the last couple of decades. Let's get specific.

SONG LYRICS

I am going to establish a benchmark that says that the lyrics in folk ballads of the early twentieth century chronicled activities of the environment in which they were composed. In other words, in the chicken and egg scenario, the songs reflected the events of time and did not "cause" the events of time. My assertion is based partly on the categorization of songs of that era. Those categories included Outlaws & Bad-men, Penitentiaries & Chain Gangs, Railroads, Tramps & Hoboes, Tragedies & Disasters, The Labor Movement, and Depression songs. A sample might be Woodie Guthrie's *Dust Bowl Blues* (late 1930's):

I just blowed in, and I got them dust bowl blues,
I just blowed in, and I got them dust bowl blues,
I just blowed in, and I'll blow back out again.

I guess you've heard about ev'ry kind of blues,
I guess you've heard about ev'ry kind of blues,
But when the dust gets high, you can't even see the sky.

I've seen the dust so black that I couldn't see a thing,
I've seen the dust so black that I couldn't see a thing,
And the wind so cold, boy, it nearly cut your water off.

Even though Woodie was known later in life for sweeping social commentary, there isn't much doubt that these lyrics were written to paint a graphic picture of life during the dust bowl era and made no presumption of global warming. I would also feel safe in stating that blowing dust has nothing to do with sex, dust getting high has nothing to do with drugs, and dust so black is not an ethnic slur. The lyrics were written in regional vernacular, most likely to add pathos to the message and not as a statement about the literacy level of the folks who lived in Kansas at the time.

It is not my intent to characterize all song lyrics of the early twentieth century as folksy. I use this example to point out how social comment was represented in song at one point in our history and how these lyrics clearly depict a snapshot of our cultural landscape. I want to use this picture to begin comparing the content of lyrics over four decades: 1930's-40, 1950's-60's, 1970's-80's, and the 1990's-2000's. As best I can, I will use lyrics that, in my opinion, depict a "clear snapshot of our cultural landscape" during each period in our history.

I will make another up-front disclaimer. I have been as diligent as possible, but there are so many lyrics in publication that time prevents me from researching the entire universe. I have also not taken the time to develop some arbitrary metric of social behavior for each song and then, applying some statistical formulae to arrive at my conclusions. In any event, you and I are having a conversation, not defending a PhD thesis.

I have not looked at classical, opera, jazz, bluegrass, rock, country, hip-hop, heavy metal, or any other genre of music in particular. I have been as meticulous as I can when searching songs that were considered "popular" during each time period. Again, I found a huge disparity in the various "Top-40" listings and made some judgment calls in filtering those lists. To be sure, I did not select any "fringe" music. I only selected "Top 10" tunes after 1970. After 1990, I only selected lyrics with clear meaning because I no longer speak the vernacular of pop music and can't imagine what some of the words mean.

First, I am going to document a few bars from sample songs that represent each decade in my curve. These songs are either highly recognizable or there is a consensus that they were "Top 40" in general air play or sales during that decade. My purpose is to get a sense for the "evolution" of the words and messages embedded in the lyrics over the last 80 years or so.

California, Here I Come (Al Jolson - 1924)
When the wintry winds starts blowing
And the snow is starting in the fall
Then my eyes went westward knowing

That's the place that i love best of all
California I've been blue
Since I've been away from you
I can't wait 'till i get blowing
Even now I'm starting in a call
California, Here I Come
Right back where I started from
where bowers of flowers
bloom in the spring
each morning at dawning
birdies sing at everything
a sunkissed miss said, "Don't be late!"
that's why I can hardly wait
open up that golden gate
California, Here I Come

'Deed I Do (Al Lentz - 1927)
Do I want you?
Oh my do i
Honey, indeed I do
Do I need you?
Oh my do i
Honey, deed I do
I'm glad that I'm the one who found you
That's why I'm always hanging around you
Do I love you?
Oh my do i
Honey, deed I do

Love For Sale (Libby Holman – 1931)
Love for sale,

Appetizing young love for sale,

Love that's still fresh and unspoiled,

Love that's only slightly soiled.

Love for sale,

Who will buy?
Who would like to sample my supply?
Who's prepared to pay the price?
For a trip to Paradise?

You're Getting to be a Habit With Me (Guy Lombardo – 1933)
I used to think your love was something that I could take or leave alone
But now I couldn't do without my supply
I need you for my own
Oh, I can't break away
I must have you every day
As regularly as coffee or tea
You've got me in your clutches
And I can't get free
You're getting to be a habit with me
Can't break it
You're getting to be a habit with me

It's a Sin to Tell a Lie (Fats Waller – 1936)
Be sure it's true when you say, "I love you!"
It's a sin to tell a lie!
Millions of hearts have been broken, yas, yas!
Just because these words were spoken: *(You know the words that were spoken? Here it is:)*
I love you, I love you, I love you, I love you! *(Ha, ha, ha!)*
Yes, but if you break my heart, I'll break your jaw, and then I'll die!
So be sure it's true,
When you say, "I love you," *(Ha, ha!)*
It's a sin to tell a lie! *(Now get on out there and tell your lie. What is it?)*

Scatterbrain – (Frankie Masters – 1939)
You're as pleasant as the morning and refreshing as the rain
Isn't it a pity that you're such a scatterbrain?
When you smile it's so delightful
When you talk it's so insane

Still its charming chatter, scatterbrain
I know I'll end up apoplectic
But there's nothing I can do
It's just the same as being in a hurricane
And though my life will be too hectic
I'm so much in love with you
Nothing else can matter
You're my darling scatterbrain

PISTOL PACKIN' MAMA (Al Dexter – 1943)
She kicked out my windshield;
She hit me over the head.
She cussed and cried and said I lied
And wished that I was dead.
Lay that pistol down, babe,
Lay that pistol down.
Pistol packin' mama,
Lay that pistol down.

Huggin' and Chalkin' (Hoagy Carmichael – 1947)
I got a gal who's mighty sweet
Big blue eyes and tiny feet
Her name is Rosabelle Magee
And she tips the scales at three-oh-three

Oh, gee, but ain't it grand to have a gal so big and fat
That when you go to hug her, you don't know where you're at
You have to take a piece of chalk in your hand
And hug a ways and chalk a mark to see where you began

SIXTY-MINUTE MAN (Billy Ward and His Dominoes – 1951)
Look a here girls I'm telling you now

They call me "Lovin' Dan"
I rock 'em, roll 'em all night long
I'm a sixty-minute man

There'll be 15 minutes of kissing
Then you'll holler "please don't stop"
There'll be 15 minutes of teasing
And 15 minutes of squeezing
And 15 minutes of blowing my top

The Naughty Lady of Shady Lane (The Ames Brothers – 1955)
The naughty lady of Shady Lane has the town in a whirl
The naughty lady of Shady Lane
Me-oh, my-oh, what a girl

You should see how she carries on
With her admirers galore
She must be givin' them quite a thrill
The way they flock to her door
She throws those come-hither glances
At every Tom, Dick, and Joe
When offered some liquid refreshment
The lady never never says "no"

The Tijuana Jail (The Kingston Trio – 1959)
We went one day about a month ago (a-ha-ha)
To have a little fu-un (a-ha) Mexico
We ended up in a gambling spot (oh yeah) a-ha-ha
Where the liquor flow-owed and the dice were hot

So here we a-are in the Tijuana Jail
Ain't got no frie-ends to go our bail

So here we'll sta-ay 'cause we can't pa-a-a-ay
Just send our ma-ail to the Tijuana Jail

Chug-A-Lug (Roger Miller – 1964)
Chug-a-lug, chug-a-lug
Make you want to holler hi-de-ho
Burns your tummy, don'tcha know
Chug-a-lug, chug-a-lug

Grape wine in a Mason jar
Homemade and brought to school
By a friend of mine 'n' after class
Me and him and this other fool decide that we'll drink up what's left
Chug-a-lug, so we helped ourself
First time for everything
Mm, my ears still ring

Mellow Yellow (Donovan – 1966)
Born high forever to fly
Wind velocity nil
Wanna high forever to fly
If you want your cup our fill
So mellow, he's so yellow
Electrical banana
Is gonna be a sudden craze
Electrical banana
Is bound to be the very next phase

Born To Be Wild (Steppenwolf – 1968)
Get your motor running
Head out on the highway
Lookin' for adventure
In whatever comes our way

Here and god are gonna' make it happen
Take the world in a love embrace
Fire all of your guns at once and
Explode into space

I like smoke and lightning
Heavy metal thunder
Racin' with the wind
And the feeling that that I'm under

HONKY TONK WOMEN (The Rolling Stones – 1969)
I met a gin-soaked, bar-room queen in Memphis
She tried to take me upstairs for a ride
She had to heave me right across shoulder
'Cause I just can't seem to drink you off my mind

It's the Honky Tonk Women
Gimme, gimme, gimme the honky tonk blues

I laid a divorcée in New York City
I had to put up some kind of a fight
The lady then she covered me with roses
She blew my nose and then she blew my mind

I Am Woman (Helen Reddy – 1972)
I am woman, hear me roar
In numbers too big to ignore
And I know too much to go back an' pretend
'cause I've heard it all before
And I've been down there on the floor
No one's ever gonna keep me down again

Oh yes I am wise
But it's wisdom born of pain
Yes, I've paid the price
But look how much I gained
If I have to, I can do anything
I am strong (strong)
I am invincible (invincible)
I am woman

Earache My Eye (Cheech & Chong - 1974)
My momma talkin' to me tryin' to tell me how to live
But I don't listen to her 'cause my head is like a sieve
My daddy, he disowned me 'cause I wear my sister's clothes
He caught me in the bathroom with a pair of pantyhose

My basketball coach, he done kicked me off the team
For wearin' high-heel sneakers and actin' like a queen

Junk Food Junkie (Larry Groce - 1976)
You know I love that organic cooking
I always ask for more
And they call me Mr. Natural
On down to the health food store
I only eat good sea salt
White sugar don't touch my lips
And my friends is always
Begging me to take them
On macrobiotic trips
Yes, they are
Oh, but at night I stake out my strongbox
That I keep under lock and key
And I take it off to my closet

Where nobody else can see
I open that door so slowly
Take a peek up north and south
Then I pull out a Hostess Twinkie
And I pop it in my mouth
Yeah, in the daytime I'm Mr. Natural
Just as healthy as I can be
But at night I'm a junk food junkie
Good lord have pity on me

Get Off (Foxy – 1978)
Ooh, ooh, ooh, ooh, ooh, ooh, ooh, ooh, ooh
Yeah
Get off
Music may ease and end all discretion
So we can get off
We keep under the sheets with two lovelys
So we can get off
Said I hope that we get the promise, ladies
And make me get off
Take it from girls with our imagination
So we can get off

ANOTHER BRICK IN THE WALL, Part II (Pink Floyd – 1980)
We don't need no education
We don't need no thought control
No dark sarcasm in the classroom
Teacher, leave them kids alone
Hey! Teacher, leave them kids alone
All in all it's just another brick in the wall
All in all you're just another brick in the wall

I Want A New Drug (Huey Lewis & The News – 1984)
I want a new drug - one that won't make me sick,
One that won't make me crash my car, or make me feel three feet thick.
I want a new drug - one that won't hurt my head,
One that won't make my mouth too dry, or make my eyes too red.

One that won't make me nervous, wonderin' what to do.
One that makes me feel like I feel when I'm with you, when I'm alone with you.

I want a new drug - one that won't spill.
One that don't cost too much, or come in a pill.
I want a new drug - one that won't go away,
One that won't keep me up all night, one that won't make me sleep all day.
One that won't make me nervous, wonderin' what to do...
I'm alone with you, baby.

Pop Life (Prince – 1985)
What's the matter with your life?
Is the poverty bringing U down?
Is the mailman jerking U 'round?
Did he put your million dollar check
In someone else's box?

Tell me, what's the matter with your world
Was it a boy when U wanted a girl? (Boy when u wanted a girl)
Don't U know straight hair ain't got no curl (No curl)
Life it ain't real funky
Unless it's got that pop
Dig it

What U putting in your nose?
Is that where all your money goes (Is that where your money goes)

The river of addiction flows
U think it's hot, but there won't be no water
When the fire blows
Dig it

Papa Don't Preach (Madonna – 1986)
Papa don't preach, I'm in trouble deep
Papa don't preach, I've been losing sleep
But I made up my mind, I'm keeping my baby, oh
I'm gonna keep my baby, mmm...

He says that he's going to marry me
We can raise a little family
Maybe we'll be all right
It's a sacrifice

But my friends keep telling me to give it up
Saying I'm too young, I ought to live it up
What I need right now is some good advice, please

BAD MEDICINE (Bon Jovi – 1988)
Your love is like bad medicine
Bad medicine is what I need
Woah, shake it up just like bad medicine
There aint no doctor that can cure my disease
Bad medicine

I aint got a fever got a permanent disease
And it'll take more than a doctor to prescribe a remedy
And I got lots of money but it isn't what I need
Gonna take more than a shot to get this poison outta me
And I got all the symptoms, count 'em 1 2 3

First you need (That's what you get for falling in love)
Then you bleed (You get a little and it's never enough)
On your knees (That's what you get for falling in love)
And now this boys addicted cause your kiss is the drug

Your love is like bad medicine
Bad medicine is what I need
Shake it up just like bad medicine
There aint no doctor that can cure my disease
Bad, bad medicine

18 AND LIFE (Skid Row – 1989)
Ricky was a young boy
He had a heart of stone
Lived 9 to 5 and worked his
Fingers to the bone

Just barely out of school
Came from the edge of town
Fought like a switchblade
So no one could take him down

Tequila in his heartbeat
His veins burned gasoline
It kept his motor runnin'
But he never kept it clean

They say he loved adventure
Ricky's the wild on
He married trouble
Had a courtship with a gun

Bang, bang! Shoot 'em up
The party never ends
You can't think of dying
When the bottle's your best friend
And now it's...

O.P.P. (Naughty By Nature – 1991)

OPP, how can I explain it?
I'll take you frame by frame it
To have y'all jumpin' shall we singin' it
O is for Other, P is for People scratchin' temple
The last P...well...that's not that simple
It's sorta like another way to call a cat a kitten
It's five little letters that are missin' here
You get on occasion at the other party
As a game 'n it seems I gotta start to explainin'
Bust it
You ever had a girl and met her on a nice hello
You get her name and number and then you feelin' real mellow
You get home, wait a day, she's what you wanna know about
Then you call up and it's her girlfriend or her cousin's house
It's not a front, F to the R to the O to the N to the T
It's just her boyfriend's at her house (Boy, that's what is scary)
It's OPP, time other people's what you get it
There's no room for relationship there's just room to hit it
How many brothers out there know just what I'm gettin' at
Who thinks it's wrong 'cos I'm splittin' and co-hittin' at
Well if you do, that's OPP and you're not down with it
But if you don't, here's your membership

Baby's Got Back (Sir Mix-A-Lot – 1992)

I like big butts and I cannot lie
You other brothers can't deny
That when a girl walks in with an itty bitty waste

And a round thing in your face
You get sprung
Wanna pull up front
Cuz you notice that butt was stuffed
Deep in the jeans she's wearing
I'm hooked and I can't stop staring
Oh, baby I wanna get with ya
And take your picture
My homeboys tried to warn me
But with that butt you got
Me so horney
Ooh, all of that smooth skin
You say you wanna get in my benz
Well use me use me cuz you aint that average groupie

Fantastic Voyage (Coolio – 1994)
Come on y'all let's take a ride
Don't you say shit just get inside
It's time to take your ass on another kind of trip
coz you can't have the hop if you don't have the hip
grab your gat with the extra clip and,
close your eyes and hit the switch
We're going to a place where everybody kick it
kick it, kick it, yeah... that's the ticket
ain't no bloodin', ain't no cripin'
ain't no punk-ass nigga's set trippin'
everybody's got a stack and it ain't no crack
and it really don't matter if you're white or black, I
wanna take you there like the Staple Singers
put something in the tank and I know that I can bring ya
If you can't take the heat get yo' ass out the kitchen
we're on a mission

I Got 5 On It (Luniz – 1995)

Player, give me some brew an I might just chill,
but I'm the type that like to light another joint
Like Cypress Hill
I'm steal doobies spit loogies when I puff on it,
I got some bucks on it, but it ain't enuff on it
go get the S-t. I-d-e-s <beer>
never the less, I'm hella Fresh,
rollin joints like a cigarette
so pass it cross the table like Ping Pong,
I'm gone, beatin my chest like King Kong,
it's on, wrap my lips around a 40,
and when it comes to get another stogie,
fools all kick in like Shinobi
no, me ain't my homie to begin with,
it's too many heads to be poppin at my friend hit it
unless you pull out the phat, crispy
five dollar bill on the real before its history
cos fools be havin the vacuum lungs,
an if you let em hit it for free,
you hellar "dum-dum-dum-dum"
I come to school with a taylor on my earlobe
avoidin all the thick teasers, skeezers, and weirdo's
I be blowin up the land like where tha bomb at?
give me two bucks,
you take a puff, and pass my bomb back
suck up the dank like a slurpy the serious bomb
will make a nigge go delirious like Eddie Murphy
I got more growin pains than Maggie
cos homies nag me,
to take the dank out of the baggie

Hit Me Off (New Edition – 1996)
1-Hit me off (oh baby, yeah)
  Hit me off (oh I like it when you)
  Hit me off
  Come on baby you drive me crazy
  Hit me off
  Freak ya like this

You got me open got me, jonin' for an episode
(Come on baby you drive me crazy)
Let's spend an hour in the shower
When it's nice and wet, I'm ready for your love

Freshmen (The Verve Pipe – 1997)
For the life of me I cannot remember
 What made us think that we were wise and we'd never compromise?
For the life of me I cannot believe we'd ever die for these sins
We were merely freshmen

My best friend took a week's vacation to forget her
His girl took a week's worth of Valium and slept
Now he's guilt stricken sobbing with his head on the floor
Thinks about her now and how he never really wept he says

I can't be held responsible
'Cause she was touching her face
I won't be held responsible
She fell in love in the first place

Jumper (Third Eye Blind – 1999)
I wish you would step back from that ledge my friend
You could cut ties with all the lies that you've been living in

And if you do not want to see me again I would understand
I would understand

The angry boy, a bit too insane, icing over a secret pain
You know you don't belong
You're the first to fight, you're way too loud
You're The flash of light, on a burial shroud
I know something's wrong
Well everyone I know has got a reason to say put the past away

Sorry, I can't take much more. Let's just fast forward and view a sampling of the Top Ten pop songs of the week of November 15, 2005 (When I started this research project).

Gold Digger – Kayne West
My psychic told me she'll have a ass like Serena
Trina, Jennifer Lopez, four kids
An I gotta take all they bad ass to show-biz
Ok get ya kids but then they got their friends
I pulled up in the Benz, they all got up in
We all went to din and then I had to pay
If you f**kin with this girl then you betta be paid
You know why
It take too much to touch her
From what I heard she got a baby by Busta
My best friend say she use to f*ck wit Usher
I don't care what none of ya'll say I still love her

My Humps (Black Eyed Peas)
What you gon? do with all that junk?
All that junk inside that trunk?
I?ma get, get, get, get, you drunk,
Get you love drunk off my hump.

What u gon? do with all that ass?
All that ass inside them jeans?
I?m a make, make, make, make you scream
Make u scream, make you scream.
Cos of my hump, my hump, my hump, my hump.
My hump, my hump, my hump, my lovely lady lumps. (Check it out)

Photograph (Nickleback)
This is where I went to school
Most of the time had better things to do
Criminal record says I broke in twice
I must've done it half a dozen times
I wonder if it's too late
Should I go back and try to graduate
Life's better now than it was back then
If I was them, I wouldn't let me in

We be Burnin (Sean Paul)
Everyday we be burnin not concernin what nobody wanna say.
We be earnin dollars turning cause we mind de pon we pay.
Worth more than gold and oil and diamonds girls we need dem everyday.
Recognize it, Pimpin as we riding.

Sugar, We're Goin Down (Fall Out Boy)
Is this more than you bargained for yet
Oh don't mind me I'm watching you two from the closet
Wishing to be the friction in your jeans
Isn't it messed up how I'm just dying to be him?
I'm just a notch in your bedpost
But you're just a line in a song
(Notch in your bedpost, but you're just a line in a song)

Shake it Off (Mariah Carey)
By the time you get this message
It's gonna be too late
So don't bother paging me
Cause I'll be on my way
See I grabbed all my diamonds and clothes
Just ask your mama she knows
You're gonna miss me, baby
Hate to say I told you so
Well at first I didn't know
But now it's clear to me
You would cheat with all your freaks
And lie compulsively
So I packed up my Louis Vuitton
Jumped in your ride and took off
You'll never ever find a girl
Who loves you more than me?

I'm sure that bastions of higher learning have spent untold grant money researching song lyrics and, no doubt, have dogmatic papers published with conclusive outcomes that are all over the map of secular and religious good and evil. Since we are not defending a Master's Thesis, my point about how song lyrics are representative of the decay in our social fabric, we'll do a quick chronology of the "message" of representative songs over the decades.

1920's - California, Here I Come – Moving to California where the weather is great.

1930's - It's a Sin to Tell a Lie - The message is in the title.

1940's - PISTOL PACKIN' MAMA - A girlfriend with a gun and a temper.

1950's - The Tijuana Jail - Thrown in jail for gambling in Mexico.

1960's - Mellow Yellow - Controversy over whether a banana peel contains a hallucinogenic drug.

1970's - I Am Woman - Declaration of the independence of women.

1980's - Papa Don't Preach - A pregnant young girl deciding to have her baby or have an abortion.

1990's - Freshmen – A young girl dies from a Valium overdose.

2000's - We be Burnin - The day in the life of a pimp.

At this point, it would be reasonable for me to make an argument and a judgment, based on the research above. In this case, however, the lyrics and my chronology will suffice as incontrovertible evidence of the rise and decay of our social conscience that is reflected in popular music of the last century.

TELEVISION

There can be no more graphic representation of our society than what is broadcast on the video airwaves and cable. Twenty-four hours a day, we can view every aspect of life in America (and the World) from food to sports and from entertainment to education. It is my assertion that the last 50 years of television is the most incontrovertible source that exists for proving our decay in social accountability as it has evolved into a mechanism for stimulating the "dark side" of human behavior. Let's first look at programming, which may be just a reflection of our society or may play a role in defining our behavior. After that, we'll look at commercials, which are absolutely designed to modify our behavior. Once again, I believe that a chronology will make my arguments for me.

First, let's take a quick look at the top-rated TV shows by decade:

1950's

Texaco Star Theatre (Comedy, variety - Milton Berle)

Arthur Godfrey's Talent Scouts (Talent search)

I Love Lucy (Sitcom)

The $64,000 Question (Quiz)

Gunsmoke (Western)

1960's

Wagon Train (Western)

The Beverly Hillbillies (Sitcom)

Bonanza (Western)

The Andy Griffith Show (Sitcom)
Rowan and Martin's Laugh-In (Comedy, Variety)

1970's
Marcus Welby, MD (Drama)
All in the Family (Sitcom)
Happy Days (Sitcom)
Laverne and Shirley (Sitcom)
60 Minutes (News)

1980's
Dallas (Drama)
Dynasty (Drama)
The Cosby Show (Sitcom)

1990's
Cheers (Sitcom)
Home Improvement (Sitcom)
Seinfeld (Sitcom)
ER (Drama)
Who Wants to be a Millionaire? (Quiz)

2000's
Survivor – Australian Outback (Extreme)
Friends (Sitcom)
CSI (Drama)
American Idol (Talent Search)

Next, let's do a quick recap of the theme and content of the shows:
1950's
Slapstick comedy (Milton Berle and I Love Lucy). A venue for launching the careers of many popular singers and comedians the next couple of decades (Arthur Godfrey). A Western where the good guys won every

time (Gunsmoke). A quiz show that rewarded genius-level knowledge, but turned out to be fraudulent ($64,000 Question).

1960's

Two more wholesome, family westerns (Wagon Train and Bonanza). Two comedy series that bordered the banal (Beverly Hillbillies and Andy Griffith). A comedy and variety show that pioneered the inclusion of political and social satire on prime time TV (Laugh-in).

1970's

A Doctor and his understudy who went about curing the medical and moral ills of our time (Marcus Welby). A news show that presented in-depth coverage of newsy items, in a judgmental format (60 Minutes). Two cute comedy shows (Happy Days and Laverne & Shirley). One comedy sitcom that challenged all previous TV rules about social, political, religious, and ethnic humor (All in the Family).

1980's

Two dramas that gave us a taste of how the wealthy exploit money and power (Dallas and Dynasty). A comedy show that actually portrayed black Americans as equals (The Cosby Show).

1990's

A sitcom that revolved around a group of misfits who spent every night in a bar (Cheers). Another sitcom that portrays an accident-prone TV show host with a dysfunctional family (Home Improvement). A third sitcom that was billed as a "show about nothing" that is also referred to as the best sitcom ever (Seinfeld). A drama that defined our paradigms about contemporary hospitals and doctors (ER). A quiz show imported from England that made $1,000,000 a new benchmark of "real" money (Who Wants to be a Millionaire).

2000's

A collection of diverse men and women live in primitive conditions for 40 days and the last one standing wins $1,000,000 (Survivor – Australian Outback). Another sitcom where contemporary life is denigrated in humorous farce (Friends). A combination of blood, guts, and special effects set in Sin City (CSI). The result of a talent search made ruthless by Reality TV (American Idol).

As with music, the degradation of social accountability in television shows pretty much speaks for itself in the brief synopses I have just presented. On Arthur Godfrey's talent scouts, the losers went home with a consolation prize. On American Idol, the losers are defamed, humiliated, and made more famous than the winners via the host of Internet sites that relive the show drama endlessly. I Love Lucy created comedy from every-day-life situations. Friends created comedy through sex and cohabitation where individuals were not accountable for their own actions, which is also (apparently) every-day-life. I can't even imagine how to create a bridge from Gunsmoke to Survivor or how they can both be at the top of the list for most watched TV programs within the same 50 year time period. If TV programs are a reflection of life, then there can be little controversy about how we have morally decayed over the last half-century. In case you are not a regular channel surfer, FOX TV seems to be taking the lead in programming innovation for the earliest part of the 21st Century. I'll leave you with brief snippets of what you are currently missing on (cable and satellite) TV. Check your local listings for times and channels.

The Simpson's - Bart sees a street sign with his name on it and the bullies entice him into stealing it. Homer makes a lame attempt at making up with Lisa. At school, Lisa goes on a rampage. Citing her father as the cause, Principal Skinner and the school psychologist talk with Homer and Marge about what they can do to prevent Lisa from growing up to hate men. Homer becomes the school safety salamander.

The War at Home - One of Hillary's friend's gets breast implants. Now Hillary wants to get implants, too. However, both Dave and Vicky won't let her do it. Dave tells Hillary that it's what on the inside that matters, not appearance.

King of the Hill - Kahn is accused of having become so assimilated that he's a "banana" -- the Asian equivalent of an "Oreo." To get back to his Laotian heritage, Kahn adopts a simple lifestyle and abandons the swimming pool his neighbors helped build.

Trading Spouses - In the two-part second season opener, we meet a woman from Massachusetts and a woman from Louisiana who swap families.

Hell's Kitchen - The celebs all arrive and begin their training. After collapsing on a garden chair, Roger is taken to the hospital because he irritated an on-going problem with his knee. He's forced to quit the competition.

Prison Break - With Lincoln's execution scheduled for the following day, Veronica comes out of hiding to contact him and gains an ally. Kellerman makes the ultimate sacrifice to preserve the conspiracy. Michael looks for a way to get Lincoln out of solitary confinement in time for the escape that evening. The inmates risk everything as they attempt their escape.

Happy viewing!

COMMERCIALS

Since the point of commercial advertisement is to entice viewers to buy a particular product or service, by definition, they are designed to influence the behavior of the viewer or listener. The logic of commercial design is that the more impact the commercial has on the potential customer, the more product will be sold. This impact can be achieved in the form of a jingle we remember, a visual image that stays with us, a compelling sales pitch, outrage, sensationalism, comedic genius, endorsement, or a host of other tools concocted by the advertising world to induce us to remember their brand name when we shop. If the content of a commercial is really successful, it will cause us to buy something we didn't have any intention of buying before we saw the commercial. Because most consumers are exposed to thousands of commercials over their viewing life, the commercials have had to become more-and-more clever and sophisticated to cause us to remember what they were peddling (We'll get into this topic in the next chapter). In the process of wearing down our resistance, it is my theory that commercials from the last decades are materially responsible, for a measurable part, of the decay of our social accountability. For example, the following is a synopsis of a commercial that ran in our local radio market (winter 2005-2006 timeframe). The quotes are my abstract of the dramatized message:

"Honey, you are speeding again." "Don't worry, dear, I have the new XYZ radar jammer. It takes out all police radar detection equipment for 3 miles. Besides, the company will pay for any tickets I get while using their radar jammer." "Well, in that case, dear, you better step on it. We don't want to be late." (Announcer) "Our radar jammer is banned in eight states. Of course we do not condone reckless-driving, however, the XYZ Company will pay any speeding tickets if your radar jammer fails to fool the police." Let's see if I got the picture painted by this commercial:

1. It's okay to speed if you don't get caught.
2. It is acceptable in 42 states to jamb police radar to prevent them from enforcing the law.
3. While we offer to pay for your fine (some restrictions apply), we failed to mention that most states have demerit points associated with traffic convictions that will likely cause your insurance rates to go up and, eventually, for your driver's license to be revoked.

It will be really challenging to find another commercial that exemplifies our current lack of social accountability as well as this one does, but I will give it my best shot as we move through the chronology of commercials over the last 50 years.

In researching the fifties through the seventies, commercials fit into a few recognizable categories. The first is "cute, catchy, and clever." These usually had a jingle or song that resonated with the target demographic, causing the customer to recall the catch phrase when shopping. They include:

Alka Seltzer's winning series of "Plop. Plop. Fizz. Fizz. Oh what a relief it is!"; "Mama Mia, that's a spicy meat-a ball!"; "I can't believe I ate the whole thing."; "Try it, you'll like it." Then there is: "Ladies, please don't squeeze the Charmin!"; "See the USA in your Chevrolet. America is asking you to call."; "Does she...or doesn't she?" (Clairol Shampoo), "Look Ma, no cavities!" (Crest Toothpaste); "Leggo my Eggo."; "Put a tiger in your tank." (Esso); "Quality is Job 1." (Ford); "Pardon me. Do you have any Grey Poupon?"; "When you care enough to send the very best." (Hallmark, Copyrighted in 1934!); "There's always room for Jell-O"; "Finger lickin' good." (KFC); "Is it Live or is it Memorex?"; "M&Ms melt in your mouth, not in your hand."; "Oh, I'd love to be an Oscar Mayer Weiner."; "Me and my RC"; "You can be sure if it's a Westinghouse"; "Ring around the collar" (Wisk Detergent).

In my opinion, this "cute, catchy, and clever" approach to product marketing engaged the consumer while they maintained a reasonable standard of taste and became part of Americana. Most of these products are still prominent in their markets.

The next category of commercials from this genre appealed to the macho-man mentality. They were in vogue until we were made aware of the detrimental effects of tobacco and that alcoholism was a national disease. They include: "I'd walk a mile for a Camel."; "Taste the high country" (Coors); "All my men wear English Leather, or they wear nothing at all."; "From the Land of Sky Blue Waters" (Hamm's Beer); (L.S.M.F.T.) -"Lucky Strike/Means Fine Tobacco"; "The Marlboro Man"; "Why don't you pick me up and smoke me sometime?" (Muriel Cigars); "Wherever particular people congregate." (Pall Mall Cigarettes); "Schaefer is the one beer to have when you're having more than one."; "You only go around once in this life, so you have to grab for all the gusto you can get." (Schlitz Beer); "Are you ready to Tanqueray?"; "I'd rather fight than switch." (Tareyton Cigarettes); "Should a gentleman offer a Tiparillo to a lady?"; "No wonder so many doctors now smoke and recommend King-Size Viceroys."

Since smoking and drinking were encouraged by the military as mechanisms for the soldiers to unwind and put the horrors of war behind them when they had time to relax, these commercials simply played to the existing market. The booze and butt peddlers have since been accused of targeting their wares for adolescents to continue smoking and drinking in their parent's footsteps.

As we move forward in the timeline, cigarette ads have disappeared from TV and radio and beer commercials have changed their focus from the gusto life to the football-on-the-couch crowd. While many of these commercials are also part of Americana, they have disappeared into history, just as movies that depict nearly everyone smoking and drinking[11].

About the most risqué commercials I found prior to the 1980's were cross-your-heart bras, 18 hour girdles, dismiss disposable douche, and the Noxzema shave cream commercial that suggests that we "Take it off. Take it all off." The most morally radical one may be the Lays commercial where we were encouraged to buy two bags and hide one for ourselves, perhaps, because of the omnipresent Frito Bandito. A real stretch might also include "How about a nice Hawaiian Punch?" While these commercials may be inane, they certainly did not glorify rampant moral decay.

---

11   I have tried watching a couple of episodes of HBO's latest despicable behavior series, True Detective. I turned it off not so much because of the ludicrous plot but at least half of each show is watching Matthew McConaughey chain smoke and guzzle beer. If he is just smoking for the part, he's destined to get lung cancer before the series is over.

The third category of commercials I have labeled are "excessive and extreme." My first recollection of commercials crossing this line was the Isuzu commercials that were launched in 1986. Their spokesman, "Joe Isuzu," was a pathological liar who made absurd representations of the capabilities of vehicles. One commercial spot found Joe Isuzu saying "and if I'm lying, may lightning hit my mother." He then proceeded to say that their 4X4 Isuzu Trooper could carry a "symphony orchestra" or "hold every book in the Library of Congress." The end was predictable. His poor mother standing beside the car went up in an exploding puff of smoke. Don't misunderstand; I thought they were incredibly clever, humorous, and effective. They did, however, create a phenomenon within social settings where many people mimicked his deceitful techniques to the degree that they became fashionably acceptable. At least in my circle of friends, these commercials elicited tacit approval for using deception as a sales tool. This is not my idea of rampant decay of social accountability, just a noticeable first step in the erosion process.

By the early 1990's, Cheer detergent stained a cloth with blueberries, drove a steamroller over it, placed it in a bowl with ice water, and then pulled out a clean cloth within the space of a 30 second commercial. In small print, there was a flash on the screen that stated "12 minutes of agitation." Anacin was running ads claiming their pill was better for pain relief than aspirin or Bufferin because it contains two ingredients. The second ingredient was antihistamine that had nothing to do with pain relief. In 1994, Playboy aired commercials peddling Playboy subscriptions and bonuses (that were actually soft porn) as suggested Christmas presents. Maalox produced their "Maalox Moment" series of commercials that suggested overeating spicy foods was fashionable behavior as long as you had their antacid with you. A Nickelodeon commercial for the Rugrats animated TV show boasted that you would not just see cute kids playing, but you would be treated every week to "greed, revenge, and deceit."

Once again, I must revisit the chicken and egg scenario. In the 1980's, the National Association of Broadcasters Television Code (U.S. v. NAB 1982) was repealed. In 1985, the Federal Communications Commission's deceptive advertising policies were eliminated. Were these two events triggered by the media's pressure to use more provocative advertising or did the repeal of the statutes open the floodgates to using sex, deception, and abhorrent behavior to sell products? No matter. We have arrived, current day, at a plateau of media advertising that actively encourages egregious behavior as a stimulus for buying a particular product or service. Here are a few examples:

A Harley-Davidson commercial includes a number of scenes of couples arriving at the female's residence after a date. The guy is driving an automobile and the women make a hasty retreat while the man is left wanting for a good-night kiss, or more. The final scene shows a Harley parked outside a residence. From inside, the moans and groans of passionate romance can be heard. The tag line is "It's better on a Harley." A Cadillac commercial dramatizes a young woman sneaking out of the house in the middle of the night. Her father discovers her missing and heads out to intercept her journey. Meanwhile, she picks up her lover and speed off into the night, until they encounter "dad" blocking their escape route. His reason for tracking them down is revealed when he states, "I don't care about you running away. Just take your mother's car (not my new Cadillac convertible)."

As I attempt to describe these commercial spots in words, I can't help but visualize the production meetings in which these story lines are hatched, scripted, and approved. Can you imagine getting paid well to build a story board for the next commercial and presenting it as a brilliant sales campaign to executives at your client's company?

Two jocks are standing outside a cabin in the woods. The preppy has a pure-bred border collie at his heals. The redneck has a mangy mutt beside him. When commanded, the collie runs to a cooler and retrieves a Bud Light Beer for his master. When commanded, the mutt jumps up and bites the preppy in the crotch, causing the bottle of beer to be ejected into the waiting hands of the redneck.

Since beer commercials appear to be setting the bar for repugnant behavior, a Miller Light commercial should be included in my rogue's gallery. Miller Light ran a series of commercials that depict debates about whether the reason for drinking their brew was its good taste or because it was less filling. As this masterpiece unfolds, two attractive women are sitting at a restaurant and the debate about great taste or less filling begins. The argument deteriorates into a boxing and wrestling match. The contest makes a stop in a public fountain where onlookers are witness to the two women tearing each other's clothes off until they are clad only in bras and panties. The brawl grows more ferocious as the combatants wind up duking-it-out in a trough of mud. The camera then pans to two jocks and two females in a bar, somehow watching this event on TV. The jocks are pumped, while the women appear horrified at what they are watching. The fighters then act out an ending to the battle, suggested by the jocks, where the two women arise from the muck, look each other in the eye, embrace, and passionately kiss. The entire premise of the commercial is implausible, antisocial, and homosexual, but somehow designed to sell beer.

The last stop on my commercial journey takes us to the state in which I reside. I live in Northern Nevada, where convention and visitor authorities have undertaken an image makeover that promotes Reno/Tahoe as a destination for outdoor recreation (skiing, snowboarding, water sports, and golf). It's good to lose our old moniker as the divorce capital of America (Yes. I do belong to the Virginia City Chamber of Commerce).

Over the years, however, I have been witness to the southern part of the State moving in the opposite direction. For some 30 years, Las Vegas was encouraging families to visit their city as a vacation destination for adults and children. In the recent past, that campaign has been replaced by Las Vegas reverting to its Sin City reputation and encouraging adults to visit for the sole purpose of partaking in adult entertainment (gaming, clubbing, and adult shows). A mainstay of several cable-TV channels has become programs about how to make the most of your Vegas vacation by learning how to gamble, where to stay, what to see, and how to score comps (complimentary meals, shows, and hotel rooms). These channels also frequently broadcast the newest craze, Texas Hold-em Poker tournaments. As an aside (or complement) to the central theme of this book, it is astounding to me that Texas Hold-em Poker has become one of the fastest growing "sports" of the new millennium. If you haven't been witness to this phenomenon and having a hard time visualizing how poker can be a spectator sport, the producers have designed the programming to be a combination of the viewer participating in the tournament (by being able to see the player's cards) and by exploiting the personalities of the star players. The "fun" shows feature has-been TV stars playing poker for their favorite charities. Somehow, watching aging entertainers like Gabe Kaplan (Welcome Back Kotter) and John Corbett (Northern Exposure) chide each other while attempting to prevail at the poker table has become entertaining to TV viewers. In contrast, the shows that feature the "real" tournament players, spend their non-table time focusing on the sordid careers and antics of former school teachers and computer nerds

Chapter 7 - The Media: Our mirrors or our teachers?

who have found their calling in Texas Hold-em Poker. Outside a casino environment, this collection of "champions" might otherwise be mistaken for social outcasts and misfits. You see, the key to success for this particular brand of poker is a mixture of mathematical skills complemented by an effective strategy for deceit. Successful bluffing is what separates a "player" from a winner. The more-clever you are at causing the other players to fold their hands as a result of your deceptive skills, the more likely you are to walk away with a million-dollar prize. I can't wait for this sport to grow more popular and be played in Madison Square Garden, with a half-time show starring Janet Jackson. But I digress.

The reason for spotlighting Las Vegas is their long-running series of commercials based on the theme that "What happens in Vegas, stays in Vegas." While the spots are produced with a modicum of class, the underlying message is always that you can come to Las Vegas and live out fantasies that are not condoned elsewhere in the world and whatever deviant behavior you can enjoy, will never be revealed to the folks back home. More vividly than any beer commercial, this ad campaign clearly establishes Clark County, Nevada as a Fantasy Island that anyone can access at will and be totally unaccountable for what happens during their stay. Coincidentally, there has been an outbreak of TV dramas based in Las Vegas, plus an HBO "documentary" series entitled Cat House. This series depicts life at the Moonlight Bunny Ranch, a legal Nevada brothel (which is about 30 miles south of Reno). If those who are tempted by the "What Happens in Vegas Stays in Vegas" commercials and also watch Cat House, they may be expecting all of their fantasies to be fulfilled on the Las Vegas Strip. Sorry Charlie. Bordellos are only legal in the smaller counties of Nevada and you have to drive an hour from Las Vegas to reach the closest one. If you look for Air Force Amy in a Vegas casino, you might wind up in jail for soliciting prostitution. Take heart, if you can't make it to Vegas any time soon, you can purchase WHiVSiV clothing online and the "Pimps and Ho's" board game is available on the Moonlight Bunny Ranch web site. Oh, I almost forgot. The cable series Gigolos is set in Las Vegas. It gives "credibility" to women who seek professional escorts to fulfill their wildest dreams, while creating role models for young men with washboard abs to explore a very lucrative occupation in servicing tourists with lots of money.

Holy Guacamole! I have reached the end of this chapter and not dealt with topics like the newly legalized prescription medicine commercials that have led us to becoming drug addicts for every ailment from stoop shoulder to limp penises. I've missed the host of ads that promise more successful love lives by wearing the sponsors' clothes or eating at the right restaurant chain. We did not delve into how legalization of commercials by lawyers has (or hasn't) contributed to frivolous law suits. I neglected to include a lengthy discourse about the influence of traditional news media versus the "fair and balanced" news media. I haven't even mentioned commercials aimed at children which may be influencing their underdeveloped value system, their lack of respect for authority, their promiscuous behavior, and their obesity. I've yet touched on political advertising, which epitomizes how to lie effectively without consequences. Stay tuned. Perhaps I will cover these topics in my sequel book, *Life in the American Communes of 2020*.

Oh yes. Of the six questions I posed at the beginning of the chapter, the only one I have a definitive answer for is about Puff the Magic Dragon. I have met Peter Yarrow (Peter Paul and Mary) and he insists he wrote it as a children's song. Since the group has a history of performing children's songs, I feel pretty confident about his assertion.

And a final thought for us to ponder: For at least the last 50 years, global perception of life in the USA has been derived from our written, film, TV, and broadcast media. Depending on the decade, my ham radio friends around the globe would want me to tell them more about those who created our national image. These "American Icons" include John Wayne, Dobie Gillis, Jim Rockford, Perry Mason, Archie Bunker, JR Ewing, Pamela Anderson, Hawkeye, and Rosanne. In film and TV, every corner of the earth has access to images of our plush homes, cars, cities, toys, pastimes, and clothing. In countries where the leaders were (are) dictatorial or socialistic, these images were (are) considered contraband and (are) kept from the general public to avoid rebellion against their highly controlled lives. In more open countries, these images have both contributed to emigration to America and have set benchmarks for the lifestyles that most citizens wanted for themselves. In the best scenarios, we have encouraged democratic and capitalistic lifestyles to grow and flourish. In the worst scenarios, we have opened our kimonos to the world and our insatiable appetite for "stuff" has caused governments to find any opportunity to export lower-priced goods for our ravenous consumption, while their own citizens enjoyed the benefits of the technology we mostly invented. Let's be real. The exports coming from China are in response to their open access to our needs. With the money they make from our trade imbalance, cities like Chunking have gone from poverty to world-class in less than 15 years.

Based on my conclusions, let's now paint some plausible scenarios for today or the immediate future. Imagine that meetings are taking place in the inner sanctums of Muslim war lords, Communist leaders in Asia, Socialist dictators in South America, and Economic Development Councils in India. They all have access to our satellite TV channels. They have staff watching our sitcoms, reality shows, and commercials. Other staff are surfing the Internet while listening to our music on their smart phone downloads. Perhaps they have done analyses similar to the ones I've done in this book on how the fabric of American society has decayed from Ozzie and Harriet to Hollywood Hillbillies and Moonshiners.

Someone in the inner-circle has also been reviewing their copies of the communist playbook, the Rise and Fall of the Roman Empire, The Rise and Fall of the Third Reich, Leadership Styles of Attila the Hun, and Guerilla Marketing. The ongoing debate among civilized leadership may sound like the following: "The Americans are following the fall of Rome. If we are to prosper, we had better exploit free trade and provide the US with as much of the decadent products and services they show us on TV before they implode. We can easily transfer their wealth and technology to our country because of the open access. We can also learn how to help our citizens avoid this level of moral decay from their commercials and TV shows."

In the strategy sessions of those who hate the US for one reason or another, their leaders may be debating the following: "Nikita Khrushchev was right. From what we are seeing on TV, if we just leave the Americans alone, they will self-destruct within a generation. Their TV shows glorify war, drugs, and decadent behavior. As long as there is a market for pornography and cocaine, we should make as much money from selling these products as possible. War with America is a form of population control, so we should not fear their armies. Their history shows that once we surrender, they will pay us even more money to be their friends."

Next time you watch Scarface and Al Pacino's character bury his head in a mountain of cocaine, think of these scenarios. When Tony Soprano whacked someone or told one of his pathological lies on the deck of his yacht, Stugats (which means "big-testicles" in Italian), think how a third-rate dictator might perceive life in the USA. The next time you see yuppies drinking late in an Internet café, envision how an Indian strategist might be plotting their next technology transfer from Sunnyvale to Hyderabad. The next time you fire-up your tablet computer, think about the elevated lifestyle you have provided to the Taiwanese that built the hardware you are using, while your neighbor is being laid-off.

Scary, ain't it? A ray of hope can be found in the news that, in 2005, John Wayne was still the most popular actor.

CHAPTER 8
THE TERRIBLE TEENS

The stroke of midnight, 1/1/2013, ushered the official beginning of the teen years of this century. I find the term "terrible teens" to be spot-on a definition of what is happening in our businesses, society, and culture as this epoch unfolds. In Chapter 5, I discuss the rise and fall of our society in great depth. This chapter is a treatise on what has happened since we bottomed-out at the end of the last century. In that century, we were busy with the "great war" and did not have time for the terrible teens. Now, we have all the time in the world to indulge our insatiable appetite for abhorrent business conduct and antisocial behavior.

The terrible teens is derived from a confluence of puberty and self-awareness that is triggered at approximately 13 years of age. The physical manifestations of bodily changes in boys and girls and the urge to reproduce are innate to our species and occur for most in this period.

These changes are inextricably connected to the discovery for the uniqueness of self. That is, we become aware for the need to be recognized for our individuality and develop a unique identity. Paradoxically, we often do that by dressing and styling our hair to correspond to the latest avant-garde fad.

Deciding if we wanted to be in the in-crowd or the out-crowd was dictated by personal preference or environment, but each of us had some manifestation of the terrible teens. That phenomenon was brought to fruition when we began to expand our knowledge base from Huck Finn and Lois Lane to social awareness and political realities.

We began to understand how unfair it was that we were of the age of reproduction, but society frowned on us freely engaging in its joys (and responsibility). Our parents spent money freely, had credit cards, and drank alcohol, but we were "too young" for such foolishness. We were forced into structured schoolwork, homework, and report cards that made our behavior public record. Most had some sort of stipend allowance that was never enough to procure the goods and services that "others" had.

The "pressure of suppression" was inevitably built to the point of exploding into one of the manifestations of the terrible teens. Rebellion against authority was a common outcome. Some worked in academic approaches to solving the problems of the ages with their peers. Others needed more overt actions against the parental, school, and civil authorities in the form of protest and obstinacy. Some embraced mind-altering substances to deal with the overwhelming pressures of their subversive environment. A few turned to attempting to change their world through violence and causing pain to the oppressors.

During the last century, there was a universal constant of societal reality that said one eventually had to grow out of the terrible teens, get a job, become accountable, and raise a family of your own. There were, of course, fringes where a few could become terrible adults and live off the public dole or derive income from criminal activities. They were in the minority and their lifestyles were seldom sustainable.

During the terrible teens of this century, we have turned the bell curve of the 20th century into the beginning of a sinusoidal wave that, as in physical science of alternating current, must move at the same amplitude on the negative side of the datum line, before returning to zero and beginning another cycle on the positive side. I am reasonably certain that I will not be around to prove the theorem[12], however, the physics of wave theory hasn't failed us yet in the scientific community and history proves that we tend to repeat human behavior in foreseeable cycles.

My theory that we are now on the "negative" side of the curve is based on the preponderance of evidence that those who believe in personal accountability and being a productive member of a society, that leaves a better world for their children, are now in the minority. Our affluence has exponentially spawned the need for more affluence. It has given us time to become aware of the chasm between the rich and the poor. Technology has made the world microseconds apart, rather than oceans apart, affording us examples of global dysfunctional behavior seldom observed before

---

12   The sine wave or sinusoid is a mathematical curve that describes a smooth repetitive oscillation.

in real-time. We are inundated with masterfully crafted propaganda in every nuance of our daily lives, creating information overload for those who are not prepared to analyze such quantities of data in determining what fact is and what mind manipulation is.

Negative, unquestionably, is the example by our politicians (at all levels) that have an inalienable right to certain perks such as food, lodging, and healthcare that is provided by the government for all who want it. Entitlement programs for everyone have replaced social programs of disaster relief for those who are accountable for themselves, but have been the victim of some tragedy beyond their control.

Ninety-nine weeks of unemployment created a new segment of our society. In that period of time, we can condition ourselves to live on that stipend at a level that is socially acceptable among other victims of capitalism. The more who join the 99 weeker's, the less spending in the marketplace. The less spending, the more unemployed! The more unemployed, the more who become unemployable; thus, the seeds of socialism are sown.

Nowadays, the terrible teens syndrome not only includes entitlement to a lifestyle that includes healthcare and a smart phone, but also ignores the reality that there are not enough resources available to sustain the growing demand. The age limit for being a terrible teen has been raised to the extremes of an expanding human lifespan.

The national credit card has had its limit removed and the conscience that used to compel us to spend less than we make has been suspended. As in wave physics, we are, apparently, going to have to reach the absolute pit of human greed before we can begin the upward curve to get back to the zero axis. This is not political rhetoric; it is basic science and math[13].

Why have I included this tirade? Because they are the people you are hiring. They either are, or will be, integral to your entrepreneurial or corporate success or doom. Many of my consulting clients continually ask me "Why don't these people get it?", "Why do I have to tell them how to do everything?", and "Why can't I motivate them to do better?" Of these and many more questions later, my favorite is "Why can't I hire my replacement?"

The reason is that we are attempting to impose our values and beliefs on individuals whose fundamental value system is different from ours as

---

13 Politics aside, The Affordable Care Act cannot pass the most basic actuarial assessment of how insurance must be structured to be effective and affordable for the consumer and feasible to administer by the underwriter.

we are from Indonesian tribal members who have never been off their island. Worst case scenario: it is now possible to be an "adult" who is second or third generation of a family that has never known a lifestyle other than that of public assistance. The concept of "working for a living" is completely alien to them, or, at the very least, an abstraction. They may have never known an actual "family" environment, being raised by one parent or another or some other relative or foster family.

On the other end of the potential workforce spectrum are those who have never wanted anything and have no concept of money, value, risk, reward, or the old adage of "paying your dues." Creating a nurturing environment of collaborative effort where everyone is accountable for their own work is becoming increasingly difficult (as if it were ever easy). Those who are reading this book likely wake up with their game face on and hit the workplace running, without a thought as to why we are motivated to work hard and achieve outcomes we are proud of. That self-motivation is no longer indigenous in our workforce. We are just as likely to have coworkers who just "show up" to put in the minimal effort required to collect a paycheck (which is necessary to pay their smart-phone bill and the lease on their new BMW).

While there has always been a chasm in corporate America between the highly motivated and those who just "have a job," work ethic is now rampant and has become an acceptable norm. Being personally accountable and being accountable to your peers is no longer the norm. Instead of coworkers challenging each other to grow, they invent alliances for mediocrity[14].

Take heart. We will soon begin exposing you to proven methods for turning around apathy and creating a healthy learning community in your organization.

---

14   I see we have a new term in our legal system. Affluenza. It is a disease of the offspring of rich parents who have no value system because they have never wanted for anything. A recent Texas case where this term was dramatized, got the defendant 10 years' probation and rehab, although he admitted to killing four people while he was driving drunk.

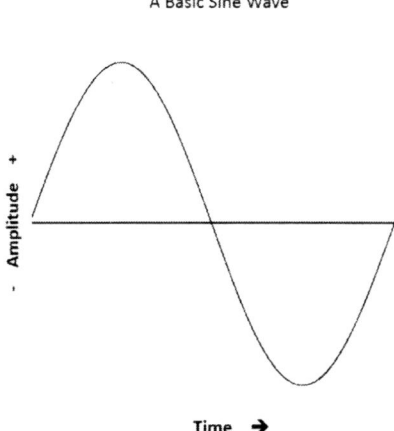

The Basic Sine Wave

PROVING THE THEOREM

In Chapter 7, I used the history of our music and television shows in the 20th century as metrics to plot the downward spiral of our social conscience and our collective moral fiber. Since 2001, I have been exposed to a virtual cacophony of television programs to prove that the sine wave was predictably continuing its negative phase of the cycle.

I do not have the stomach to look at current music lyrics. My previous examples will have to stand on their own merit. I also don't have the constitution to watch ANY broadcast TV, even in the spirit of book research. The plot lines for these shows have become so inane that it insults anyone with a fifth grade education and has the sensitivity of a boulder. These shows are also punctuated with magnificently choreographed commercial messages, with equally banal content, that is designed to subliminally entice the audience to use their credit cards or food stamps on products that will change their lives. In the last decade, you weren't manly if you didn't smoke Marlboro cigarettes. Now, you need Shaq O'Neil's body lotion to "man-up." Neither of these messages challenges me to saddle-up and head to Wal-Mart, but I am clearly not the demographic for the later.

In my satellite TV channel surfing, I am able to (mostly) avoid stations with commercials. Recently, however, I clicked on a movie containing full frontal nudity, R-rated sex acts, and graphic violence that was interrupted by a beer commercial, a tire commercial, and then, to my great surprise, a lengthy commercial message for Trojan personal stimulation devices (I presume the batteries are included). The commercial was tastefully produced and exposed me to another technology intended to enhance our lives with cutting-edge electronics and fractional horsepower motors.

I have recorded several seasons of the programs I will chronicle, so that I am certain, beyond reasonable doubt, that I have accurately captured the intended message of each writer and producer communicated to ME. I freely admit to being perversely entertained by each of them. At the same time, I can assure you that none of them caused behavior modifications to your humble narrator. Once again, this data is provided for you to study what your employees are obsessed with at night and what state of mind they may be each morning.

Let's begin with the dramatization of Roman History.

SPARTACUS

The telling of Roman history is never complete without a chapter on their apparent refinement of greed, avarice, and personal aggrandizement to its lowest level of sloth, which lead to the fall of the empire. History books tell of the games in the arena leading to the death of the weaker gladiator. They describe opulent orgies of communal sex fed by excessive food and drink. They portray the caste system as the very rich victimizing the very poor and rationalizing it with the approval of their gods.

Roman legend portrays Spartacus as the slave transformed into the penultimate invincible gladiator. Spartacus inevitably uses his charisma and fighting skills to break free from his role in the arena and begin a revolt among the slaves to become free men. He is also ostensibly on a quest of revenge for the death of his wife at the hands of the Romans. Each episode is another phase of the rise of Spartacus as the leader of the slave war against the establishment.

At this point, it sounds like a dramatization of Roman history that is worthy of a series on PBS. To make it palatable to the 21st century audience, however, we are witness to some of the most creative special effects Hollywood has to offer, punctuated by sex scenes that used to be seen only on adult videos. Each episode depicts varying degrees of full-frontal male and female nudity, male-female and male-male rape, heterosexual and homosexual acts, lavish orgies, and an occasional male actor sporting an erect male member. These are not fleeting glances, but prolonged and graphic scenes, which I did not know were in the acceptable column in the TV-MA ratings guidebook.

The battle scenes rival the best sci-fi graphics for the appearance of reality. The creator of the show has assembled the very best in slow-motion photography and computer animation to create the most realistic pictorial representation of beheadings, swords through the head and torso, limbs being severed and flying through the air, and warriors dying excruciating deaths covered in burning pitch. Blood spurts in glorious profusion from severed arteries. Gouged-out eyes appear to be true empty sockets of gore.

The men are as handsome and perfectly proportioned as the best gigolo would be. The women are not only stunning, with magnificent femininity, but they have eyes of fire and compelling features that make them unique and unforgettable among beautiful women.

The acting is superb. The few scenes, where the players have their clothes on and are not wielding swords, have compelling dialogue of stilted Roman speech patterns from actors who have pseudo-British accents. The moments of non-violence are typically the characters revealing their devious plots of subversion and treachery to gain power, wealth, or to promote the cause of freeing the slaves. There is even a compelling scene of compassion between two of the female leads, once enemies, now friends, that ends with one performing an involuntary caesarian section on the other and the women and baby all dying spectacular deaths.

For those viewers who do not watch overt pornography because you have to pay extra for it, Spartacus is about as close as you can come to X rated sex and violence for the price of a subscription to Starz. You can even get an "app" for your smart phone or tablet to watch the shows at a convenient time for you. I can't wait for it to be available as an adult game on X-Box. It will likely have a new manifestation of the classic "joystick" in its control features.

Finally, as a public service, Spartacus is also a popular slot machine game at your favorite casino. While the bonus-round-graphics do not include violence and nudity, they inevitably lead to relieving you of your coin.

Does art mimic life or life mimic art? In this case, the questions are: 1) Is the pretext of dramatizing the fall or the Roman Empire grounds for moving to the next level of graphic violence and gratuitous sex? And 2) If Ancient Rome is the archetype for civilizations, how many years are we from the fall of our Republic?

HOMELAND

Once again, the producers have assembled some of the most talented actors, perfectly cast, this time to play compelling roles of CIA agents, Marines, and terrorists. The main character was a POW imprisoned by al-Qaeda, who lived in a hole where he was tortured for years. His family believed he was dead, only to be freed by his captors and returned to the USA to a heroes' welcome in the first season. Reunion with a wife and family, who believed you were dead, makes for high drama and clever plots.

The female lead is a CIA agent who is convinced that he was turned by his radical Muslim captors. Her mission is to uncover the nefarious reasons for his release. She has obtained unverifiable evidence of him becoming a traitor and cannot convince her bosses at the CIA to allow her to conduct surveillance on the Marine hero. She becomes an obsessed rogue and works outside the law to prove her assertions.

So far, Homeland is playing out as great spy drama, while clues are being revealed that may confirm that he is brainwashed to perpetrate subversive deeds against the US Government. Government employees exceeding their Constitutional power is certainly not worthy of a headline story plot, but the plot does thicken. President Barack Obama has praised Homeland, and is also known to be a fan of the show[15].

By the end of season 1, we find that the heroine was correct and the Marine hero was, in fact, turned by his captors to exact revenge for the death of the son of the al-Qaeda leader who befriended and then released him. When a faulty suicide bomb vest and a call from his daughter thwart his attempt to blow-up himself and the Vice President of the US, the plot for season 2 is set for him to become a US Congressman and become the enemy within.

The reason I add Homeland to this chapter is because we are lulled into caring for both the traitor and the CIA agent operating outside the bounds of propriety of oath. We feel empathy for the Marine who would kill the VP, the CIA agent that lives on ADHD drugs hidden from her superiors, and even the terrorist leader whose son was killed by a US drone

---

15   TV Guide. Retrieved September 24, 2012

strike. I also include it because the subplot of the Marine and the CIA agent is to conduct a clandestine romantic affair that, again, emotes the viewer to root for the thwarted romantics[16]. While the show was praised by the critics and received numerous awards, others panned it as the most Islamophobic show on television, portraying Muslims under the light of simplistic concepts and as a monolithic, single-minded group whose only purpose is to hurt Americans. Could Homeland be a vehicle for anti-Semitic propaganda against all Muslims? At the very least it explodes terrorism into our living rooms in a plot that brings pathos and validation to illicit romantics.

Does art mimic life or life mimic art? In this case, the questions are: 1) Are most viewers seeing the plots as parable to teach the danger of radical Islam? And 2) Is Homeland a fair barometer of corruption within the CIA?

CALIFORNICATION

Glorifying sex and every other mainstream vice, this black-comedy series also moves amoral behavior into the spotlight of activities that fans of all ages need to model. The hero is a drunken novelist who, for some reason, emits pheromones that cause virtually every woman in the series to crave his sexual favors. He, of course, has not the willpower to deny any of them his manhood.

The heroine is an extraordinarily beautiful actress who mothered his adolescent daughter and continually takes the hero back into her life and bed, regardless of his transgressions. As the seasons evolve, so do the characters depicting alcohol and drug dependency, most acts of sex from the Kama Sutra and a supporting cast of self-abusers and sex addicts. There is hardly a moment in each half-hour session where drugs, sex, and obsessive behavior are not being lionized. Each character has some major character flaw that is glorified while they are represented as a typical group of friends living in the Venice, CA area. Even the opening cut on the theme video has the hero pointing to a poster for a "freak show."

Finally, as each season closes, the anti-social behavior patterns are unceremoniously expunged, but set up another mind-boggling story line of sex, drugs, and rock and roll for the next season. Again, the writing and acting are superb, but the glorification of abdication of personal accountability overshadows the entertainment value.

---

16   Without giving away the entire subplot of the third series, the heroine is carrying the baby of the traitor

Does art mimic life or life mimic art? In this case, the questions are: 1) When did glorifying promiscuity, drug use, and alcohol become acceptable as the main theme of even a cable TV show? And 2) Is the behavior of the cast creating the "new norm" of middle class behavior?

DEXTER

Turning from dark comedy to dark drama, the award-winning Dexter series includes a hero who is a serial killer and a cast of villains who are also obsessed with serial murder, but the hero killer always prevails over the villain killer. The actions of the hero are justified because he witnessed the butcher-murder of his mother when he was a toddler and his ghost-father has taught him to kill only those evil perps who fit his code of vigilante justice.

In the early seasons, Dexter cleverly dispatches the bad guy of the season, while using his work as a forensic pathologist to hide his activities from everyone except for the apparition of his father, who returns from the dead at critical junctures to remind Dexter of his "code." He also keeps his kill lust secret from his good-cop sister and a number of romantic partners. When he spawns a son, the infant becomes his only mortal confidant as daddy tells the infant of his work as a serial killer, hoping to leave mental imprints on the baby to not follow in his footsteps.

As with all successful TV series, depicting antisocial behavior and abdication of personal accountability leaves the writers scrambling for plots that will give the maximum empathy for the heroes and villains in a story line that doesn't bore the viewers. This gifted crew has done the job to the accolades of the critics and the viewers.

In the 2012 season, the plot line needed elevation from good killer kills bad killer. To that end, the cop-sister discovers the murdering field of her brother so that the storyline includes the corruption of the pure-of-purpose cop to ensure her brother is not caught. What a great segue from rogue serial killing to creating impunity with a sister, who is the lieutenant of homicide, helping you cover your tracks! The final episode leaves you wanting for more. Perhaps his adolescent son will become his accomplice in preparing Dexter's kill room for his next ritual murder.

Does art mimic life or life mimic art? In this case, the questions are: 1) Does Dexter provide a new model for vigilantism? And 2) Is this the precursor to a series about likeable pedophiles?

HOUSE OF LIES

I must admit, being a management consultant for two decades, there was no way I was going to miss a TV show that portrayed the most senior members of my profession practicing our trade. The promos portrayed our lovable hero as the most cunning, two faced, conscienceless, Lothario imaginable. The first two seasons have not disappointed.

The members of the consulting firm exhibit varying degrees of psychotic behavior that utilizes their MBA degrees from the most prestigious schools, as they support complex plots to extract obscene consulting fees from each engagement. More gratifying than the fees, the "consultants" pit one abhorrent CEO against others whose greed and lack of conscience rivals Bernie Madoff.

Mixed into the plots are weekly sexual encounter that is so dysfunctional that it makes a visit to the Mustang Ranch as typical of a family outing as a day at SeaWorld. The hero also has a gay tween son who bounces back and forth between his condo and his mother's, who also uses her body and cunning to win consulting engagements.

The heroine in this plot is an up and coming female version of the hero, whose moral fabric is only as thick as her superficial beauty and less viscous as her heart of molten lava. There is hardly a good looking client or senior member of the firm who does not partake in her feminine favors. The two other guys in her "pod" have extreme interpersonal issues that make her look like Mary Poppins.

Good thing each episode is only half an hour long. I would not be able to stomach an entire hour of my profession being portrayed as the sleaziest occupation imaginable.

Does art mimic life or life mimic art? In this case, the questions are: 1) Does the series reflect the realities of the greed, avarice and corruption of corporate America? And 2) Will the MBA schools add the show script lines to their case study curriculum?

NEWSROOM

The tease for this series portrays our hero, a seasoned cable news anchor, interacting with a college coed who asks a question, in public forum, that is so naïve in political reality that sets him off in a tirade that would make a Tea Party speech writer envious. His soliloquy is so eloquent and powerful that I assumed every liberal thinker would take stock of their political beliefs and change parties.

Back at the cable news headquarters, this un-journalistic behavior causes a shakeup in the staff. The first season evolves with the dysfunctional interaction of the newsroom folk. Each episode also takes on some hallmark political or social topic that is played out by the hero in a news broadcast of journalistic excellence and remarkable disclosure of a body of evidence that leaves no room for controversy. I applaud the writers for mixing politics back and forth from liberal to conservative. It must serve to ensure that all demographic groups are drawn to the show to garner fodder to support their own political views.

The show has not received the same accolades from the critics and audiences as the others we discussed. I have included it because it depicts a plausible scenario of how the news media can distort and polarize any event to create the drama necessary to bolster their ratings. While the senior staff and the other newsroom staff maintain the appearance of propriety in confirming stories and sources, when deadlines loom, if it bleeds, it leads. Discovering the truth behind the story is not as important as being the first to press with it. I am eager for the second season to make high drama of the 2012 elections and ensuing financial controversies. Perhaps the writers can offer a plausible dramatization of how congress has not passed a budget in years, in direct contravention of the Constitution.

Does art mimic life or life mimic art? In this case, the questions are: 1) has Newsroom taken the lipstick off the pig and dramatized the real world of the current broadcast news media? And 2) Will those who are not critical thinkers absorb the plot lines as an accurate representation of the news?

SHAMELESS

I've saved the show depicting the absolute lowest moral compass for last. Shameless creates comedy from a welfare family in a Chicago slum, whose parents are the most amoral individuals, short of pedophiles and child killers. The father is a drunk and druggie whose life is devoted to contriving schemes to pay for his addictions. He is often homeless, unless he is in the midst of some scheme to extort money from a naive widow or some other unwilling halfwit. The mother has been written in and out of the series as another drug-dependent slut who has no conscience when it comes to feeding her needs.

The children range in age from toddler to twenty-something. Some are from different mothers and fathers, spawned during various drunken stupors. The oldest daughter runs the household, as best she can, in between involvements in her own dysfunctional romantic relationships, which are played out in graphic detail. One son is gay, but wants to get into West Point. Another is an academic genius that uses his talents to spite the establishment and exploit the world of illegal commerce. Another is an arsonist and fledgling criminal. The other daughter is a tween who has embraced deceit and larceny as a lifestyle, but her motivations are the welfare of the family and friends.

Some of the amazing story lines include the father having buried a relative in their front yard so he could continue to collect her Social Security check each month. They have to move the skeleton when the city needs to work on the water pipes. He also convinces the arsonist son that he had cancer in a scam to collect money from the agencies that help the terminally ill. The plot is thwarted when the agency sends his son to summer camp, instead of paying off in cash. Of course, the trip to summer camp includes the son talking to a female camp counselor to show him her breasts. One of my favorites of "How in the world did they think of that?" plot writing is when their water heater breaks. They look in the obituaries for some old person who died recently and they remove the water heater from the vacant premises. Genius!

Each character is depraved in one form or another, developed in story lines that glorify homosexual rape, teenage promiscuity, and confidence schemes that rival any contrived plot line I have ever heard. Unlike the other examples, there is absolutely no socially redeeming theme, message, or story line. It is antisocial behavior made into black comedy that appears to be a plausible lifestyle, with no attempt to separate abhorrent behavior from an entire community to the community activities of the 1950's series, Father Knows Best.

Again, it is compelling entertainment, anticipating what massively outrageous behavior the writers have created for the next episode. It is not my fear that the audience members will mimic their behavior, it is my fear that their actions will redefine socially acceptable conduct, validating our move to complete abdication of social accountability.

Does art mimic life or life mimic art? In this case, the questions are: 1) Is Shameless creating the new morality for those who find themselves in poverty? And 2) See question #1.

DOES LIFE MIMIC ART OR ART MIMIC LIFE?

In the sixties, this was often a subject of lengthy didactic debates in the coffee houses of Greenwich Village and other bastions of philosophical study. At this point in my life, I propose that the answer to the conundrum is "YES." I submit that both are true to some extent. While it is a reach to call some of the TV shows above "art," this riddle is profound in my conclusion to this chapter.

First, I offered my theory that our society, as a whole, is living out the symptoms of a teenager going through awareness, discovery of self, and puberty. I, then, used the theorem to suggest evidence that, in my observation, in 2001, we entered the negative cycle of a sinusoidal wave that was destined by physical law to bottom out at an amplitude similar to my conclusion that we peaked as a moral society in 1950.

Also in Chapter 7, I chronicled the change in social morality evidenced by music lyrics and TV shows, as they became more graphic, sexual, and scatological in content since 1900. That evidence is the bases for the following conclusions:

- Life is a landscape from which artists observe and create sensory audio and visual representations of how they see the world. Those who partake in the art, either expand their horizons through the stimulation of the work, mimic the work in their world, or find justification for their own beliefs through the art.
- Independent thinkers include art in the many dimensions of information they gather each day. It may be relaxing, stimulating, and/or educational. Just as all other data in our lives, it does not materially change our behavior by its content. Rather, it contributes to continually updating and validating our guiding principles by which we behave in our various environs.
- Those whose lives are mostly guided by sensory stimulation from others, pointing their moral compass in the direction of the latest data set observed, are those whose lives mimic the artistic stimulation. Independent thought is either a gift they do not own, or they are lazy and abdicate their value system to be manipulated by the influencers around them.

Therefore, life mimics art and art mimics Life!

Applying these conclusions just to the TV shows I used in my examples of our "Teen Art" without even discussing the effects of X-Box and Play Station games on our youth are examples of the escalation of degree of violence and abhorrent behavior. I submit that the creators, writers, and producers of these shows observed life and created dramatizations of what they observed. To garner desired revenues and ratings, they had to make the art bigger than life, farcical, and include antisocial, violent, and sexual behavior that most of us would not think of or experience on our own. After all, Buck Rogers evolved into Neil Armstrong when we used science fiction to create a model for what is possible!

Those independent thinkers, who partake in this art, can find it entertaining and stimulating without committing antisocial acts as a result of the stimulation. If, on the other hand, only a fraction of the viewers are those who mimic art and find this behavior to be a life enhancing opportunity, then our society is headed for Sodom and Gomorrah at a rate of speed and velocity that may be irreversible.

More frightening than that, to me, is that this behavior may redefine social conduct so that what I label as antisocial, pornographic, criminally stimulating, scatological, and abhorrent today will become the new benchmark of societal behavior. This proposition is not a breakthrough in creative thinking, it is a product of the pendulum that exists throughout recorded history, swinging back and forth from angelic to demonic behavior and everything in between, as the accepted societal norm.

In an episode of Star Trek, Captain Kirk was attempting to explain to an alien life form the moral fabric of our human species. To paraphrase, he submitted that humans were barbarian by nature, but we chose not to be barbaric, <u>today</u>. My takeaway is that we volitionally chose our behavior on a daily basis in accordance with the environment we are in, the stimulation presented to us, and our ability to suppress our barbarism for another day. Within our global society, we have the gamut of behavior that runs from missionaries sacrificing their lives for the betterment of others to those who endeavor to eradicate their neighbors because of a difference in theologies.

In my sine-wave hypothesis, we reached the positive peak of our potential about the time we decided to send men to the moon. In seven years, we just made up our minds and did it. Our society was thirsty for data about space, technology, and human triumph. Isaac Asimov stimulated our minds with science fiction, which was a challenge for us to make real. We had heroes who were test pilots and military leaders. Congress was comprised of representatives of their constituency, not career politicians. Graft and corruption existed primarily in organized crime and the bad guys were eventually brought to justice.

As the sine wave gains amplitude in the negative direction, our society has become so overloaded with data that we cannot separate fantasy and science fiction from any form of reality. Most of us are unable to gather raw data, analyze it for ourselves, and reach our own conclusions. Every data source on the internet or in the media has already spun their bias into raw information as they vie for ratings and search hits. We see real-time information from every continent everyday and do not have the bandwidth to process the data, so we rely on others to do it for us. It doesn't matter whether we listen to PBS or FOX News, we are being offered process data that will shock us with its graphic content.

In our downtime, we will be exposed to TV shows, video games, movies, sporting events, and other distractions that can expand our horizons or modify our behavior. Even more so today, than in the days of Hitler, Stalin, and Lenin, charismatic rhetoric and plausible theories of economic development are absorbed by more and more individuals without researching their long term effect on our globe and our societies. In 1950, we did not have credit cards. By 2050, we will not have credit cards again because our global economy based on "spending tomorrow's money today" will have completely collapsed around us.

Heroes no longer exist in the Terrible Teens, unless they are movie stars, rock stars, and athletes. Many of those are victims of their own success, who eventually self-destruct. The heroes should be the team that built and successfully landed the new Mars Rover. I'll wager that none of us knows a single name of that amazing group of scientist and visionaries, but we know the names Dexter and Shaq O'Neil and identify with their infamy.

Instead of heroes who inspired us to achieve impossible goals, we now have so-called leaders who are herding us into a common state of mediocrity. I now understand why we are not sending men to Mars in this decade and I am ashamed that my generation has created this legacy.

The takeaways from this chapter include:
- We have created a societal and moral fabric that is NOT the foundation for us achieving greatness in business and commerce.

- We are at a point in our evolution where we have forgotten the reasons that socialism and communism have never achieved sustainable outcomes.
- Our Country is founded on rising from oppression and creating entrepreneurial opportunities from all adversity thrown at us. We have forgotten this lesson almost entirely.
- We are born of craftsmen and innovators. We are morphing into models of mediocrity and apathy.
- I have spent this amount of time setting us up for the solution because our future really is as dire as I paint it if we do not get our heads out of the sand and begin a revolution to regain our greatness. I also need to reiterate that all of this setup is necessary for us to internalize WHY and HOW we got to this place and to remind you that these are the gene pool you have to interview as employees.

As they say in the infomercials, "But wait!" I'm not quite there yet.

Will The Cycle of History Follow the Laws of Physics?

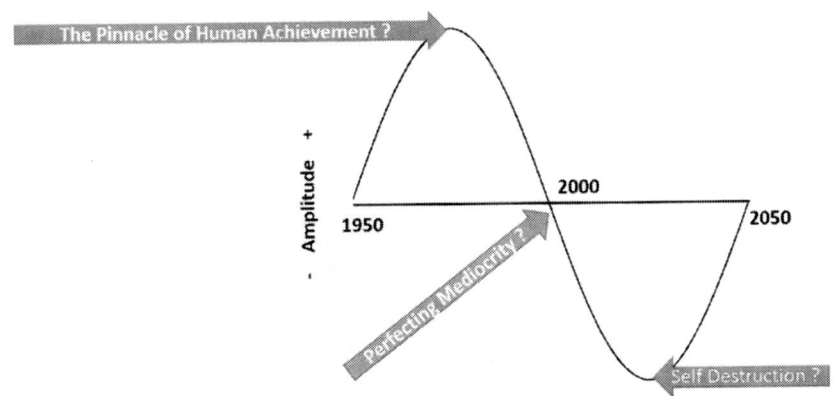

Is this Our Real Human Time Line?

See the Bell Curve on Page 59

CHAPTER 9
LIVING LA VITA VICARIOUS

I still remember the days when my father sold pots and pans door to door. Part of his sales pitch was how much time a housewife could save with the "waterless" cookware he was peddling. So much of advertising in the 50's was about the latest labor saving devices or how quickly we could wash the dishes, vacuum the floors, and do the laundry if we purchased the newest products. As an adolescent, I really didn't understand (and really didn't care) about labor saving devices because I had no concept of the value of time. As is common in adolescents and teens, I hadn't become aware that time was not a limitless resource.

As I constructed my Achievement to Apathy curve for the 20th Century, I realized that it wasn't all that long ago that we were an agrarian culture and nearly every hour of every day was consumed with some sort of work, maintenance, or very slow travel between home and where we needed to be. The Industrial Revolution moved populations closer together, provided mechanical labor, shortened travel time, and left the workingman with actual downtime that could be budgeted for recreation. The more we automated the mundane, the more time we could spend with elective activities and the term "pastime" came into our vocabularies. Early in the twentieth century, pastimes tended to be social activities that were opportunities for relaxation because most folks' daily routines still included a significant amount of physical exertion.

As recently as the 1950's, I recall Monday nights at the movies, bingo at church, and a weekend trip to the beach with significant pastimes for my family. We actually planned for weeks for the adventurous 30 mile trip to a day-long picnic at the lake.

From mid-century on, cars went faster and automated appliances and machines did much of our physical work for us, making our chores and our jobs more sedentary than they have ever been in the history of mankind.

Not only were we freeing up time, we had become less active that we had to manufacture physical work around the house to keep from becoming stout and lethargic. This new phenomena was manifest as the "The Weekend Handyman." An offshoot of my father having more time was that and much of *my* free time was consumed with involuntary labor helping to build retaining walls, laying patio bricks, and continually building bars, tables, and shelves in the basement. The more TV commercials we watched, the more labor saving devices we bought - freeing up more time to build the do-it-yourself projects suggested by other commercials. Instead of milking the cows, we now spent our time changing the wallpaper in the den and the linoleum in the kitchen every time a new color scheme became available, and that was before color TV! As I strongly suggested in the last chapter, TV commercials have highly influenced our behavior over the last 50 years and I lived through the dawn of this phenomena.

During the first half of the century, we had limited exposure to what others did for pastime activities when most of the visual input we had come from Life, Look, and National Geographic magazines. Once we were able to watch the Wide World of Sports on TV, our horizons were greatly stretched. The more leisure time we had, the more TV channels came on the air and the more we learned about activities that were previously unknown to us. By the eighties, we had plenty of free time and hundred channels of cable TV that exposed us to every imaginable activity on the planet. The law of supply and demand was in its glory as producers searched the world to discover more and more unusual topics to present to us, so their market share would grow from providing more spectacular events than the competition.

Until the sixties, the universe of activities shared by most of the population of our country may have commonly consisted of football, baseball, basketball, bicycle riding in the summer, and sledding or ice skating in the winter. School activities might have included swimming and track. Today, we have added soccer, golf, tennis, lacrosse, skating, skateboarding, wave surfing, wave sailing, jet-skiing, water skiing, river tubing, snowboarding, snow skiing, snowmobiling, mountain biking, moto-cross, camping, equestrian, repelling, rock-climbing, dodge-ball, paint-ball, shooting, archery, and an activity I just discovered another one called

"Fusion." The definition I found for Fusion is (sic) a fun and supportive environment where youth can share, explore, and celebrate the richness and complexity of a mixed heritage. The program facilitates self-discovery through creative and interactive activities such as visual and performance art, outdoor education, community service-learning, field trips, and student leadership activities. I'll borrow another Dirty Harry maxim to sum up my feelings about "Fusion" with his sarcastic use of "Marvelous" as a response to such arcane information.

When I was young, baseball and ham radio consumed all of my free time. When my sons were in school, between the two, they played baseball, football, basketball, performed in the band, and rode their bikes for transportation. Present day, my grandson is involved with baseball, football, soccer, basketball, bicycling, dirt biking, tubing, and video games. All three generations have had about the same amount of time to devote to extracurricular activity, however, our appetites appear to have increased based on what is currently offered to us.

My point thus far is that we have spent decades creating less and less work for ourselves so we can become involved in more and more pastime activities. The more activities we create, the shorter our attention spans have become and the larger our appetites have grown for more active and passive stimuli. The more challenging the activities become, the more we must push the envelope to experience the *Rush* I described in Chapter 5. The more extreme sports become, the more our thirst grows for pushing the limits of human endurance. Boxing and wrestling have been upstaged by XFC Extreme Fighting, where pretty much anything goes as long as it inflicts pain on your opponent. Pretty soon, we will fill arenas with men fighting lions and tigers to the death, or has that been done before?

We've become so skilled at efficiently growing, harvesting, and processing our foods that we can unconsciously consume ten times the amount of calories we need on a daily basis to sustain our activity level (I heard a quote some years back that it will take the human body another 1,000 years to evolve to where it can assimilate all of the ingredients in a Big Mac with Cheese). We spend billions of dollars on foods, supplements, and drugs to cure the obesity that the last few generations worked so hard to create with labor saving technology. We have become so accomplished in replacing muscle with tools that we now must set aside some of our pastime to exercise our bodies. We actually pay to go to health clubs to burn surplus calories. What is wrong with this picture?

For those who regularly partake in any aerobic activities, visual stimulation at least comes from nature and an animated environment. For many of us, the limit of our physical exertion may be ten hours a day of pushing a computer mouse or joystick. For this growing demographic, the visual world exists mostly on a cathode ray tube, plasma, or LCD display (TV's or computer monitors). For future generations, it is possible that three-dimensional activity may only be experienced inside a virtual-reality helmet instead of a football or motorcycle helmet. If there is any hope that violent video games are fantasy and not stimulation to act out their sadistic themes, it may be the argument that the fanatical game-players haven't got enough muscle tone to actually implement the abhorrent behavior they act out on the Game-boy. That brings us to the main theme of this chapter, La Vita Vicarious.

Before we start, let's again do a quick recap of the last century. We first evolved from an agrarian and physically challenging society, to an industrialized one. Daily life in the early part of the millennium still required significant physical activity, but we worked hard and perfected the automobile and other labor-saving tools that allowed us to become more relaxed and, ultimately, more sedentary. Since we created planned free-time, we created activities to replace work, called pastimes. Instead of farming for a living, we now garden for fun. Instead of building our houses and barns, we build bars and bookshelves in the den. Instead of physical exercise as part of our daily work, we create physical activities for play. As technology spun off more and more devices to help us and entertain us, and more and more information to guide us, we found out our free time has become occupied with assimilating data on the TV and on the computer. We now "watch" from the den many of the physical activities we used to take part in or we occupy a seat at an air-conditioned stadium or theatre or in front of our Smart Phone and Tablet. We've been so incredibly effective at distributing limitless quantities of visual stimulation that we have created a national epidemic of dulled-senses, leading to more need for more stimulating data to fill our ever more titillated appetites. Hence I have coined the term "La Vita Vicarious," or living life through the activities of others instead of actively participating in it.

We have arrived at an evolutionary point in civilization where we must fill a growing need to stimulate our senses at higher and higher levels of *Rush*. What gave us a thrill last week is now old hat. The chase scenes, bloody carnage, and collapsing buildings in last month's movies are now just a benchmark for the next epic to exceed, coming to a theater near you next Friday. In early movies, when a bad guy was shot, the actor grabbed the supposed-wound and fell to the ground moaning. There was never any blood, guts, or gore. Then, Hollywood invented squibs under clothing that actually made gunshots or stabbings look like a bloody wound. Then we advanced special effects to where body parts appear to be ripped from the torso and blood rhythmically spurts from severed arteries. In the latest TV masterpieces provided by the TV series Crime Scene Investigations, we now see dramatizations (in slow motion) of bullets penetrating a victim's chest cavity, rupturing the liver, causing green bile to fill the abdominal cavity, leading to a slow and painful death. The same TV series has provided us a primer on what type of insects grow in dead bodies at what times, facilitating the investigators' ability to establish accurate times of death for homicide victims. I can hardly wait for the new season in the fall to see how the producers push the envelope of "blood and gore" visual effects.

Instead of reading a mystery novel and imagining a murder in your mind's eye, you can now experience the act in 3D, 1080p High Definition, missing only the smell and feel of the actual event (and those effects are available at live events such as wrestling matches and auto/motorcycle races). Movies, TV, and the computer have provided us with such graphic images and sounds of almost any type of human and animal activity imaginable that we can essentially save the energy and expense of physically experiencing life and live vicariously through the activities portrayed on a screen. Since bodies at rest tend to stay at rest unless acted upon by an outside stimulus (according to Isaac Newton), we should soon atrophy and assume the body contour of a Laz-y-Boy recliner. The more prices drop on 70" Smart TV sets and the more shopping you can do on the Internet, the less need there will be to leave home for any reason, except maybe for cardiac or gastric bypass surgery. Yes, we do still fill football stadiums, but we now do it after the tailgating party. Thanks to another labor saving device, we can have our new portable kitchen installed in the back of the pickup and we can cook and watch TV while we drink beer with our best buddies and get "in the mood" for the kickoff.

I again ask if television portrays life or does it create new realities for us? Is La Vita Vicarious a product of consumer demand or is it a product of marketing genius? I also feel the need to defend my continual focus on TV and computers as being huge influences on our lives and the La Vita phenomenon. Check out the Sunday sale papers. They are predominantly promoting home entertainment products, not sporting goods. There are cottage industries blossoming that specialize in installing home theatre systems. You can purchase High Def TV sets at computer stores and you can buy computers at department stores. Most computers now have DVD players and TV tuners so you can watch TV while you are on the computer. Many vehicles have DVD players to entertain the passengers (and further distract the drivers). You can watch TV or surf the Internet on your cell phone. Looking around my place, we have six TV sets and four computers; only two of us live here. I just realized we also have a small battery powered TV we take when we go to our secluded weekends on the beach in Mendocino. Egad! The last time we went to that Bed & Breakfast, I even took my laptop and projector and we screened the digital pictures we took each day! I am part of the problem! I know I am not an "addict," however, because we do not watch sitcoms or extreme TV. I assuage my conscience with the fact that I predominantly watch The Food Channel, Discovery Channel, National Geographic Channel and, of course, mob movies (it's a Sicilian thing that is a byproduct of my heritage). As I decide whether I am in TV denial or not, let's take a look at how the 21st Century TV producers are satiating our need for La Vita Vicarious.

We'll start out slowly with some program highlights from the Extreme Sports Channel (All listings are from Internet program guides. I couldn't make this stuff up):

gen://ex (that is the title of the show, not a Web address): Extreme's lifestyle mag hooks up with skate/snow prodigy, Shaun White, checks out some tattoo art, and looks at new mountain bike video 'Anti-Gravity'.

X-Terrain: Magazine show featuring adrenaline fuelled sports from around the world from mountain biking to skateboarding.

O'Neill Pro Freestyle 2005: Sick snowboarding contest from featuring superpipe, slopestyle, and back-country competitions. Plus a huge freestyle rail session.

P.I.G.: The ultimate game of one-upmanship featuring the world's top motocross riders battling to be the last man standing (riding).

Hardcore Candy: All girl magazine show featuring snowboarding action from the Roxy Chicken Jam and a road-trip with some of the hottest surf girls.

Now let's sample an assortment, across broadcast and cable networks, of picks and pans from the "Reality TV" program listings:

Amish in the City (UPN): This show follows Amish youth in a rite of passage known as Rumspringa. Amish people at the age of 16 leave their homes and communities and test the outside world and enjoy the things forbidden in the Amish world. They must then determine whether or not to return to the Amish life. This show follows five Amish and five non-Amish living together in this process.

Beauty and the Geek (WB): After the eight new women choose a partner, the contestants move into the mansion, and quickly begin to bond. Josh finds himself filled with so much anxiety and insecurity that he cannot bear to sleep in his room with Cher but instead spends the night in the closet while Tristin reveals her own insecurities about wearing a swimsuit.

Con (Comedy Central): Skyler Stone is a con artist and can get anything he wants for free. Why? Because he's a con man. His new show is about showing you how you can live the con lifestyle, too!

Damage Control (MTV): An all-American teenager is left home alone for the weekend while hidden cameras follow his or her every move. The parents watch from next door with the host, trying to guess just what their son or daughter might do next. Each time they guess right, they win $1,000. When their parents return, each time the teenager confesses to what happened they win more money.

Fire Me, Please (CBS): Two people begin new jobs in hopes of getting fired by 3p.m. the same day. The contestant fired closest to the 3p.m. deadline wins $25,000.

Girls v. Boys (Noggin): Six teens face-off to prove who rules. Cameras follow each cast of six very different teens as they compete in the ultimate battle of the sexes! From episode seven, in the competition "Sea Joust", Demian and Krystal each show how aggressive they can really be. In "Worker's Comp", everyone's wit and charm is put to the test.

I Want to be a Hilton (NBC): Kathy Hilton, onetime actress and mother of Paris, stars in this reality-competition series. In the season finale, the final two contestants receive pronunciation training, and give a final speech that Kathy Hilton will use to decide the winner.

Kept (VH1): Model Jerry Hall, chooses 12 men to compete for the opportunity to live her lavish rock 'n' roll lifestyle as a "kept" man. The lucky competitors will share in her life of luxury, accompanying her to star studded events, rock 'n' roll parties, and socializing with her friends and family in London.

Meet the Barkers (MTV): Travis Barker, the heavily tattooed, mohawked former drummer of Blink-182 has found true love with a once Miss USA turned TV actress Shanna Moakler and they've invited the whole world to watch as they begin their bizarre life together. From the first moment they appear on-screen, we see just how free spirited and over-the-top Travis and Shanna really are. From hangovers and Playboy parties to Thanksgiving dinners and their child's first birthday party, Travis and Shanna are living the rock n' roll fantasy while simultaneously adjusting to parenthood.

Pet Psychic (Animal Planet): Sonya Fitzpatrick talks to animals using the mind. Her readings are real, done with real animals.

Queer Eye for the Straight Guy (Bravo): Five gay men are out to make over the world - one straight guy at a time.

Sheer Dallas (TLC): Get a glimpse behind the scenes for an insider's look at the larger-than-life characters who make up the social strata of this first-class city.

The Smoking Gun TV (CTV): A show about legal matters concerning celebrities, stupid people, and weird cases, done in a humorous way.

The Surreal Life (VH1): Trapped without transportation, cell phones, or personal assistants, has-been celebrities must interact with each other, share bedrooms and bathrooms, do household chores, go grocery shopping, and prepare meals together. The cameras never stop rolling, so the power struggles and personality clashes are all captured on film.

Wildboyz (MTV2): An action/adventure show spearheaded by its two stars of low moral caliber. In each episode, the boys travel afar to exotic lands as America's foremost ambassadors of absurd goodwill, engaging in close cultural encounters with a diversity of dangerous wildlife and native peoples.

Okay, let's "amp it up" for our finale, and switch channels to Spike TV. This latest channel replaced The Nashville Network and is advertised as being "The First Channel for Men" (Its program guide is quite dissimilar to WE, the Women's Entertainment Channel, however). Here are some highlights:

MXC: "MXC" (formerly known as "Most Extreme Elimination Challenge") is the ultimate in reality sports, where two teams of contestants are both physically and mentally challenged and eliminated through crazy games. In episode # 60, Las Vegas vs. Sesame Street, Big Bird takes on big bets when *MXC* mixes together two of the most opposites in America: The children's playground of *Sesame Street* and the adult playground of *Sin City*.

Super Agent: Nine sports agents compete to be selected as agent for USC All American defensive lineman Shaun Cody. One agent will be eliminated in each episode as they fail to make it through the tasks assigned to them to Cody's satisfaction.

The Ultimate Fighter: Accomplished athletes, disciplined in wide range of mixed martial arts that include wrestling, boxing, judo, Jiu-Jitsu, karate, and kickboxing compete for the dream, the hope, and the will to fight in an Ultimate Fighting Championship event, which plays out as, essentially, "anything goes" except homicide.

Maximum Exposure: Scantily clad sky-divers, drunken flirting, cheating spouses, and drug sting.

Real TV: Barfly boa constrictor; train wreck; hummingbirds.

Well, that should just about do it for our research on reality and extreme TV. I don't know about you, but my senses are overloaded with compelling invitations to enrich my life experiences by watching these shows. With just a small sampling of what is available to us each viewing day, I am feeling compelled to give up my outside activities and get the *Rush* by living vicariously through TV personalities. Ya think?

If you do decide to actively partake in these offerings, let me recommend that you maintain a level of sanity by peppering your schedule with regular episodes of Penn and Teller - Bullshit (Showtime). With biting sarcasm, these two Vegas entertainers will provide you with a sanity check about what is really going on in the world of our extremes and realities.

Given just the quantity of this fare, I am convinced that La Vita Vicarious is a national epidemic. I must also conclude that the fight for higher ratings will continue to make the producers of these shows more inclined to redefine "extreme" on a weekly basis, until we are watching actual killings, raw sex, and pedophilia on cable TV. It is clear to me that TV is providing this programming because we have an appetite for the

Rush that is almost insatiable. It is still unclear in my mind whether the content of these shows mimics life or establishes a pattern that causes us to behave more and more abhorrently. Let's keep this question open as we move on to the next topic in our journey to It WAS Rocket Science.

Oh, I almost forgot to remind us that these shows are broadcast to our friends, competitors, and enemies worldwide as the stereotypical images of life in these United States. In time, they will even be able to buy this book from Amazon.com!

The point of all of this is that these are actions and the images burned into your employees' brains just before they fell asleep last night!

CHAPTER 10
SUCCESSFULLY AVOIDING PERSONAL ACCOUNTABILITY

It's amazing how experiential and anecdotal data for this book presents itself to me nearly every day. Recently, the owner of an "upscale adult heath club" was giving me a tour of his facility, in hopes of selling me a membership. We walked into the men's locker room and he was both visibly upset and embarrassed that there were soiled towels thrown about on the floor. Some weren't even near the receptacle provided. He apologized for the clutter, picked up the towels, placed them in the bin, and we moved on. He made the comment to me, "I can't believe that our members are so thoughtless. I'll bet they don't do that at home." I related a story to him that may have been my earliest awareness of our societal movement toward perfecting mediocrity.

As the manager of the electronics manufacturing company, in about 1983, I inherited a nephew of one of the owners. He was offered to me as someone who could do mechanical assembly work, drill holes, and paint metal. I quickly discovered that he was a born-and-bred Texas redneck who dipped snuff and chewed tobacco among his other bad habits. Many of my best friends were rednecks. The difference with this young man was that most other native Texans had a sense of personal pride and accountability. Some of my employees smoked in the work areas and his dipping would not have been a problem, except he would spit out the remnants just about anywhere he could find a receptacle (trash can, soda can, etc.). Moreover, his aim was terrible, so he left trails wherever he was working. I mentioned several times to him that his habit was unsanitary, but "Uncle" was given to a pinch between his cheeks and gums, so my young man felt invincible and uninspired to be accountable for his habit. The women who worked on the production line called themselves

The Pasadena Princesses and they were well experienced in the habits of redneck men. As tolerant as they were, a group of them came to me one day and said that even they could no longer stand the sight and smell of used snuff everywhere they turned and what was I going to do about it. I was skilled in the power of peer pressure, so we had an ad-hoc meeting to which everyone was invited. We had a lively discussion about tobacco and cigarette smoking and decided to eliminate smoking from the work place. They then turned to the redneck and started on him about his alternative-tobacco habit. At the height of the debate, the senior Princess (who outweighed him two-to-one) got directly in his face and demanded, "You don't spit tobacco on the floor at home, do you?" His response was, "Damn right. That's why I got an old-lady for." The room fell quiet. Not usually at a loss for words, we collectively had no response. I don't think I will ever forget how intensely his illiterately-worded reply resonated with me. First, I was dumfounded that this tried-and-true challenge of social behavior had failed for the first time. Secondly, I had no idea that his wife, who appeared to be a reasonably intelligent young lady, was volitionally living in an environment where she dutifully cleaned his nicotine slobber off the living room floor. I found out later that this was just one of her assigned duties, which also included cleaning the fish he caught, and gutting and dressing the deer he hunted. Oh yes, she was also tasked to keep a certain level of beer inventory in the house at all times.

We resolved the snuff issue by getting him to agree to using it only in his work area, keeping the remnants in one used soda can, and not partaking anywhere else besides his pickup truck. The Princesses agreed to only smoke outside the building, which led to later discord about abuse of break times, but that's another story. I came away from the event with my first benchmark of the depth of irresponsibility of some of the citizens of Gen-X.

During the same time period, I was an assistant coach for my youngest son's little league baseball team. As a coach, I had encountered lack of discipline, respect, and accountability, but only from a few of his peers. I dismissed their behavior as an anomaly, because these were the same kids whose parents never attended games or team events (In Texas, many parents live for sports events and push their kids to be superstars, so these absentee-parents were social outcasts by definition). The head coach was a strict disciplinarian and my role was to play good-cop by encouraging the team to use peer pressure to motivate the troublemakers to conform to the rules, at least on the ball field. Since we had a history of producing championship teams, the leadership and the peer pressure kept their behavior in check and kept me insulated from the lack of discipline and accountability these studs exhibited off the field and at school. I learned later, that their need to stay on a winning team was the only motivation they had to behave with any social accountability, anywhere in their lives. It wasn't until years later, in the early days of my consulting career, that I made the connection between how the dysfunctional behavior I observed on the ball field manifested itself in the workplace and in the daily lives of those who were destined for mediocrity. I also learned, years later, that our ball field had been built on top of one of the worst toxic waste sites in the Country. I'm still trying to make a correlation between toxic waste and the unusually high number of championship teams that played on that field.

DEFINITION - BABY BOOMERS[17]

A baby boomer is a person who was born during the demographic Post-World War II baby boom and who grew up during the period between 1946 and 1964. [1] The term "baby boomer" is sometimes used in a cultural context. Therefore, it is impossible to achieve broad consensus of a precise definition, even within a given territory. Different groups, organizations, individuals, and scholars may have widely varying opinions on what constitutes a baby boomer, both technically and culturally. Ascribing universal attributes to a broad generation is difficult, and some observers believe that it is inherently impossible. Nonetheless, many people have attempted to determine the broad cultural similarities and historical impact of the generation. Thus, the term has gained widespread popular usage.

United States birth rate (births per 1000 population), the blue segment from 1946 to 1964 is the postwar baby boom.

In general, baby boomers are associated with a rejection or redefinition of traditional values; however, many commentators have disputed the extent of that rejection, noting the widespread continuity of values with older and younger generations. In Europe and North America, boomers are widely associated with privilege, as many grew up in a time of affluence. As a group, they were the healthiest and wealthiest generation to that time and amongst the first to grow up genuinely expecting the world to improve with time.

17   Excerpted from Wikipedia

One feature of Boomers was that they tended to think of themselves as a special generation, very different from those that had come before. In the 1960s, as the relatively large numbers of young people became teenagers and young adults, they, and those around them, created a very specific rhetoric around their cohort, and the change they were bringing about. This rhetoric had an important impact in the self-perceptions of the boomers, as well as their tendency to define the world in terms of generations, which was a relatively new phenomenon.

The baby boom has been described variously as a "shockwave" and as "the pig in the python[18]." By the sheer force of its numbers, the boomers were a demographic bulge which remodeled society as it passed through it.

DEFINITION - GENERATION X (THE 13TH GENERATION)[19]

Those associated with Generation X have cultural perspectives and political experiences that were shaped by series of events. This includes the end of the Cold War, the fall of the Berlin Wall, Vietnam, the late-60's space race, the 1973 oil crisis, the 1979 energy crisis, the early 1980s recession, the Chernobyl disaster, Black Monday, and the savings and loan crisis, both of which preceded the early 1990s recession. Generation X saw the introduction of the home computer, the beginning growth of video game era, cable television, and the Internet. Other attributions include the U.S. urban decay, the AIDS epidemic, the War on Drugs, the Space Shuttle Challenger disaster, the Iran hostage crisis, Iran-Contra Affair, Operation Desert Storm, the Dot-com bubble, alternative rock, and the global influence of the hip hop culture and music genre. They are often called the MTV Generation. Pertinent to a non-partisan study on the 2008 U.S. Presidential election, the Population Reference Bureau, a demographic research organization based in Washington, D.C., cited Generation X birth years as falling between 1965-1982.

In American cinema, directors Kevin Smith, Richard Linklater, and Todd Solondz have been called Generation X filmmakers. Smith is most known for his View Askewniverse films, the flagship film being *Clerks*, which focused on a pair of bored, twenty-something convenience store clerks in New Jersey circa 1994. Linklater's Slacker similarly explored young adult characters who were more interested in philosophizing than

---

18 A sharp statistical increase represented as a bulge in an otherwise level pattern, used especially with reference to the baby-boom generation

19 Excerpted from Wikipedia

settling with a long-term career and family. Solondz'. *Welcome To The Dollhouse* touched upon themes of school bullying, school violence, teen drug use, peer pressure, and broken or dysfunctional families, mostly set in a junior high school environment during the early to mid-1990s.

When compared with previous generations, Generation X represents a more heterogeneous generation, exhibiting great variety of diversity in aspects such as race, class, religion, ethnicity, and sexual orientation.

Prior to the 1960's, our society was all about accountability. Immigration had produced more workers than available jobs and most employees of that era were diligent about their performance in the workplace. This accountability was taught to their children and, even though the job-to-worker ratio became more level, pride in a job well done was the norm. As suburbia came about, this pride was manifested in how each homeowner worked to keep his home well maintained, the grass trimmed, the flower beds attractive, and the community a reflection of their newly found affluence. Saturday in Levittown was a cacophony of lawn mowers and the sound of garden hoses, as we mowed, edged, and washed the family car. Sunday was dress-up in the morning for church and some family event for the afternoon. These scenarios were pretty common among the baby-boomers who I have interviewed about the fifties and early sixties.

In Chapter 5, we talked about the positive attributes of FDR's New Deal. There were also some unintended outcomes of that program. In 1935, The Social Security Act was passed as an "insurance" program for retirees to ensure that they could provide for themselves during their golden years so that they would not be a burden to their families and communities. The Act also included a national welfare system to assist needy children. This provision turned out to be an invitation for social abuse of the program. Prior to the Social Security Act, communities had provided "welfare" benefits at a local level, to help out-of-work citizens get on their feet and find new employment. The community looked out for the needs of their children with care and encouragement to help them rise above whatever tragedies had befallen their families. The federal program wasn't administered by concerned community elders; it was quickly exploited as a "dole" instead of temporary assistance. It became such a scam that unemployed men found it practical to abandon their families so that the wife could get Social Security benefits for herself and the children. For the next 60 years, we discovered more and more dodges for social accountability and invented many more schemes for abusing welfare. It wasn't long until inner-cities had become communes of welfare "families" that invented clever ploys to replace work and accountability entirely with social programs. For those who had the least amount of self-respect, women kept having children out of wedlock, so they could collect more and more benefits, while the absentee-fathers found ways of using their kid's food stamps for booze, tobacco, and gambling. After a generation or two, the communities evolved a mutated group of citizens born into a society that never lived anywhere besides a welfare state. Their birthright included free money and social handouts as an entitlement, not a transition or temporary insurance. Even though the Welfare Reform Act of 1996 was passed to return the country to use of Social Security as a temporary bridge to help individuals rejoin the workforce, the "entitlement" mentality of at least one entire generation had left an indelible stain on the fiber of our society.

Even for Americans who never saw a welfare check or food stamps, as memories of the Depression and the World Wars started to fade, we began to lose some of the values taught by our parents. There's nothing new about selective amnesia. Humankind has historically forgotten about "bad" history, as the events of the moment occupy our thoughts. If we could develop the skill of continually reminding ourselves of the wisdom afforded to us by disasters of the past, we would have a fighting chance at eliminating wars and revolutions. Unfortunately, the wisdom learned from prior social traumas did not affect our behavior as we moved into the late sixties. Those who only witnessed the welfare state in the media (or maybe in their towns) developed a certain amount of resentment for those who lived on the public dole as a lifestyle. Eventually, images of hundreds of families abusing welfare took its toll on how the middle class approached social accountability.

For those who actually worked at productive jobs, discretionary time and money was another new phenomenon that blossomed during this era. Our self-imposed need to make more money created the two-income family. We were then introduced to the first wide-spread symptoms of couples abdicating their duties as parents, as they worked longer and harder in order to support their "keep up with the Jones" lifestyles. When I was in high school, both of my parents worked to support their desired lifestyle. Fortunately, my sisters and I had developed a value system based on fear, and we mostly stayed out of trouble as latchkey kids. For some, however, instead of being serenaded by the buzz of a lawn mower, they were unsupervised and free to explore the buzz of marijuana and the untended liquor cabinet. The unaccountable youngsters I coached in baseball were merely the product of the evolutionary social decay begun by absentee-parents of the sixties.

Even though my parents may have worked as many hours as their parents, the motivation was different. My grandparents worked hard just to keep bread on the table and shoes on their feet. My parents worked hard for their own comforts and for their retirement years. With all due respect, for the post-depression-era parents, I do concede that there was also a motivation to ensure that the baby boomer generation went to college and would not have to work as hard as our parents or grandparents to make an acceptable living. While my dad repeated the mantra often about how he was working hard to send us to college, I wound up paying for my own education and I turned out to have a value system based on the discipline he taught us and on reaping the rewards of my own hard work. Many of my contemporaries were given the same mantra from their parents about providing them the means to go to college, except they filtered the message to mean that their parents would provide endless resources so they wouldn't have to work at all (rather than work "as hard"). For the first time in our history, we were matriculating high school and college students who were unaware that they could not instantaneously move from their parents' home into the same lifestyle provided to them for by mom and dad. How rude was that?

For those who were prepared and self-motivated, they moved out and worked hard to build careers that would allow them to enjoy the lifestyle they grew to enjoy at the expense of their parents. For most of them, renting an apartment and buying a starter-home were character building steps of growth as they worked their way up the food chain. For those who did not anticipate life in the real world, they were shocked at how few amenities they could afford with entry-level jobs. This group either learned from their past mistakes or eventually grew up or they are still living less-than-fulfilling lives within the welfare state. For those who spent their teen years exploring drugs, sex, and rock & roll, facing the real world was "totally" intolerable and they created a new reality based on blaming their problems on everyone else. Instead of focusing on an education that would lead to a bountiful career, many of my peers dabbled in existentialism, the projects of Sartre, and the promises of socialism. They learned that rebellion against all established norms was their birthright, regardless of the price-tag their iconoclasm levied on society. Encouraged by Timothy Leary to tune in, turn on, and drop out, they evolved a new culture of codependent and needy individuals who, at best, exploited friends and social programs to survive and, at worst, turned to illegal activities to sustain their chaos. When they left the campuses of liberal teachings, they discovered how detached they were with societal reality. Prospective employers were not impressed with their credentials in sociology and their work experience protecting spotted owls. Disdain for the "establishment" was another of their creeds that often ended job interviews. I still see many of these individualists, with shoulder length gray hair and peace symbols on their trench coats, walking the streets of Reno, pushing shopping carts containing their worldly possessions.

While previous generations of indigent and homeless citizens came as a result of personal or natural tragedies, this new generation had evolved as a product of apathy and lack of social values. Many migrated west, where they found refuge with other kindred spirits who had discovered California's limitless welfare handouts. The hippie generation set up communal living models that created a new social order based on "letting it all hang out" and doing whatever feels good. How could we have known that a few mutant offspring of well-meaning parents (who provided for their children's every need) would spawn a national epidemic of irresponsible and antisocial behavior? I am not a psychologist or sociologist, but it would appear that it is embedded in the human genome for some of us to depart from our cultural norms and training and follow a path of least resistance that leads to total abdication of accountability and self-respect.

Add to the welfare state, the irresponsibility created by misguided parents, the divisiveness formed by the Vietnam Conflict, and racial unrest. We have cemented the foundation for the downfall of social accountability for the rest of the millennium. By the 1980's, the pride and self-motivation brought to Ellis Island by our forefathers had morphed into a society where lack of self-respect and values based on entitlement were allowed to exist, unchallenged and unashamed. Those who practice unaccountability and social irresponsibility are not the majority, but the fact that they have our tacit permission to coexist while every other sector of society has diminished the values of our entire population.

My three sons are now in their 40's and they share a high set of values that we have modeled for them. They work hard and the two who have children, are raising our grandsons responsibly. With all that said, their daily lives are preoccupied with sports and pastimes. The fruits of their labors satiate their need for La Vita Vicarious, their versions of the *Rush*. There isn't an ounce of entrepreneurial spirit among them. While they work hard, there isn't a visible passion for raising standards, quality levels, or accountability within their business lives. Work is a means to an end. That end is more selfish than the first half of the 20th Century. They don't spend any time working to improve their communities or to share values and tribal knowledge with the next generation. I'm not saying that they lack social accountability, I'm just pointing out that they may not realize that they, and their contemporaries, are building a legacy of perfecting mediocrity instead of continually advancing the state of humanity. Sad, methinks, but then they probably think I am an anachronism because I am continually evolving new businesses and writing more books after my 65th birthday.

Since I've defined the context for how our downhill curve of social accountability has evolved from the welfare mentality, the love generation, the absentee parents, and from Vietnam and racial strife, let's know look at the effect all of these have had on our 21st century predisposition for avoiding accountability.

Have you watched the typical 21st Century family in public places like the mall? Since we no longer believe in corporal punishment or damaging the child's psyche with harsh words, children run amok while the parents repeatedly threaten them with "time outs" or just ignore them. The kids have learned that there aren't any real consequences to behaving badly, so they do. Study these children and adolescents on your next trip to a store. They have no social skills and push their way into, around, and past anyone in their way. There are no courtesies shown to adults. The art of holding a door open for the next person is virtually nonexistent. Handling and mishandling merchandise that was not paid for is common. In restaurants, there is no attempt by parents to have youngsters maintain their behavior in respect to other patrons. Between rowdy children and parents on their cell phones, decorum in a restaurant is a distant memory for me. Children in buffet environments are mostly left unattended as they push their way through lines, spill food, and use their fingers as serving spoons. While these few examples of antisocial behavior are unacceptable by themselves, they set the model for these same individuals avoiding accountability as they grow into adolescence.

Lack of accountability as a youngster not-surprisingly breeds teenagers who lack respect for the law. Go back to the driving school story in Chapter One. In several of my examples, the most basic rules of law pertaining to driving vehicles safely were routinely ignored because there are no consequences. If you recall, in one example, mom and dad did not care that their daughter was caught driving with other underage kids after curfew and paid for her to take defensive driving to avoid a conviction. This abdication of accountability came after the girl had not only been cited by the police for these infractions, but after she repeatedly broke the law by continuing her original journey to and from a party. Just think of how accountable she and her friends will be once they have their permanent drivers' licenses and can behave without conscience 24 hours a day on any public road! These are the same children whose lack of accountability allows them to grow up, spill coffee on themselves, and then sue McDonalds for negligence.

Yes, like it or not, we are all residents of the "sue nation" and there is probably no better example of our evolution to lack accountability than is provided by the current state of our legal system. Just as I believe that allowing the airing of TV commercials for prescription drugs has spawned a society of hypochondriacs and pill abusers, I believe the new generation of commercials for law firms has elevated our thirst for transferring blame for our personal (and corporate) lack of accountability to whomever we can find to sue.

Recently, one of my associates was sitting on a panel at a conference in a local hotel-casino meeting room. He inadvertently moved his chair too far backward and fell off the dais. As he recovered and placed the chair back in its proper place, he was swarmed by hotel personnel. First they wanted to know if he had injured himself. He said, only his pride. Then they wanted him to have a medical checkup and sign a series of liability waivers. He told him that it was his fault, he was fine, and there was no need for medical attention. This admission of personal accountability was apparently alien to the hotel personnel. They were very insistent that he at least sign the liability release forms. He soon discovered that tripping and falling accidents had become so common that the hotel's insurance carrier routinely paid virtually any claim made by a patron, so they were instructed to obtain liability releases immediately after any alleged incident. Upon further questioning, he was told that it was almost a weekly occurrence to have someone claim to fall on an escalator because it was nearly impossible to prove that the escalator had not malfunctioned in the moment of the alleged accident. He pursued this phenomenon and discovered that it was common knowledge that almost anyone could claim a tripping and falling accident in a public building and receive a quick insurance settlement. The insurance companies had developed an actuarial posture that is more cost effective to settle than to have to deal with one of the hordes TV-commercial lawyers who make their livings filing frivolous law suits. He and I speculated that this trip-fall-collect-money process might become the model for a new career path for third-generation welfare recipients.

If you are with me on the argument that decay in self-respect and social accountability transformed the Social Security Act into the foundation for a welfare state, then my next proposition states that frivolous law suits are a logical extension of redistributing wealth from those who have it to those who believe that they are entitled to other people's money. If I believe that I am not accountable for my actions because of how society victimized me as I was growing up (dysfunctional families, drugs, crime, etc.), then it stands to reason that I should be allowed to exact revenge on any fat-cat that I can find who might be vulnerable to being duped out of some of his or her wealth. To revisit one of the more famous examples of this, if we weren't becoming systematically devoid of accountability, how could anyone even conceive of the notion of suing McDonalds when their own clumsiness caused them to spill hot coffee in their laps? Coffee is typically served hot in this country and is most often enclosed in an insulated container. Anyone who had ever drunk coffee in their life quickly discovered the technique of sipping fresh coffee carefully to prevent burning our lips or mouth. Self-preservation taught us to handle the container of hot coffee with a high degree of care. How could anyone look someone straight in the eye and claim that their own lack of accountability was somehow the fault of the restaurant chain? My other problem is how could McDonald's exacerbate the situation by capitulating and settling the claim by paying money to the recalcitrant plaintiff? What's the learning moment here for our society?

As expected, the "lesson" motivated another individual to try a similar stunt with Wendy's, as a means of potentially collecting a six-figure settlement. A woman claimed that she found a severed fingertip in a container of Wendy's chili. The chain suffered an enormous loss of revenue and a terrible black-eye to their otherwise exemplary reputation as a result of the law suit. In this case, sanity prevailed. The plaintiff and her husband were recently convicted of planting the fingertip in the chili and both received prison sentences for fraud. Hallelujah! There is hope for us.

So how has this spiral of decaying traditions established by welfare, absentee parents and our legal systems manifested itself in the work force of the 21st century? The lament I hear repeatedly from my consulting clients is that, "We just can't hire and retain competent and motivated employees." This phenomenon is most pronounced in Nevada because we have enjoyed the lowest unemployment rate in the Country for many years. Those who do not have jobs are mostly those reasonable employers who we would not want as employees. Also, our fair State has three predominant industries with very unique cultures. In the world of gaming, casino employees who deal with money must have spotless backgrounds in order to get a work permit. Since squeaky-clean individuals are rare, casino workers who meet muster can usually get a job regardless of their productivity or work ethic. Thus, there is a continual revolving door among the casinos that are trying to find "clean" workers who also <u>want</u> to work. Mining workers live in a rather closed community of like-minded individuals. There is also a small gene-pool of candidates for these difficult jobs. Construction workers are, by design, itinerant and move where the work is. The available pool of construction workers are most often misfits, incompetents, and unreliable. What more difficult demographic could I find to discuss avoiding accountability in the workplace?

Let's begin with jobs that do not require a college degree. I want to be clear, without factory workers, construction people, store clerks, and plumbers, our Country would be in shambles. I thank my trash collection folks every time I see them. They have what many of us would consider menial and thankless jobs. Without them behaving professionally, punctually and skillfully, however, our communities would literally be a stinking mess. I have great respect for anyone who takes pride in their work, regardless, if they are changing my oil or performing brain surgery. In fact, a few weeks ago I encountered an individual that I would like to introduce to every one of my clients who complains about poor employee performance.

On a cold and snowy day in December, my sewer line backed up for the second time in a week. After calling half the plumbers in the yellow pages, I finally found one who would actually come and perform any necessary repairs within the same week. It turned out that a 20' section of sewer pipe had to be replaced. Because of the location, the plumber dug a 20' long by 4' deep trench, by hand, in the snow, wading in ankle-deep muck and waste. I stood on the porch, shielded from the weather and told him that I admired his work ethic and that I could not possibly do his job. He stopped shoveling for a moment, turned to me and said, "You know, it's just plumbing. It's what I do." What a learning moment! When he decided to become a plumber, what many of us see as nasty and laborious tasks, became just routine components of his job that had to be done with skill and tenacity. When he finished, he relayed that this was his last assignment for the day and was going home to work on his vintage Harley Davidson motorcycle for an upcoming weekend rally. He volunteered that he was paid so well for being a plumber that he could afford the luxury of his prized motorcycle and had enough time off to enjoy it. I shook his hand and returned to my office, where I spend at least 60 hours a week "just consulting." It's what I do. How come I don't have the money or the time to restore a vintage Harley? Considering the high standard I have set with my new plumber-friend, let's discuss what is happening with the other 85% of the non-degreed workforce.

Once again, current events present me with abundant examples of my thesis. Last night on the local evening news, there was a story about how some downtown citizens were being inconvenienced by two building projects, on opposite sides of the same street, causing both sidewalks to be closed for some period of time. This was requiring pedestrians to detour one city block from their normal route. The person-on-the-street interview was flabbergasted your humble narrator. The young lady, wearing the service garb of a downtown establishment, told the reporter that, because of the sidewalk closing, her boss could expect her to be late for work each day the detour was in place. The fact that the average non-exempt employee would not automatically adjust their departure time by a minute or two to allow for the detour, screamed a value set that was totally alien to me. Did she not show up for work at all when it rained or snowed?

For the rest of this chapter, my remaining conclusions about the shift from personal pride to mediocrity are derived from my consulting experiences and from recent survey data and magazine articles.

THE BOOMER WORKERS

There are still a few stalwarts who have had long-time trade careers with one or two employers. These are the workers who still believe in craftsmanship and give both extraordinary effort and loyalty to their employers. This former majority is now a minor-minority. Most trades and crafts people have been jaded by the evaporation of jobs that they thought might last for a career and by retirement plans that were under-funded or mismanaged. Those who have not been directly affected by layoffs or the disappearance of their planned pensions have been witness to the rampant behavior of employers disregarding loyalty and/or their promise of a company-funded retirement program. Most feel no loyalty from their employers, beyond the current pay period. Many believe Social Security will be bankrupt by the time they are able to collect their share. Rather than focusing their best-efforts on the needs of their potentially-temporary employer, they are trying to figure out how they can start a 401K plan at 45 to 55 years of age and, maybe, have enough money to retire at 65. That strategy has also turned to folly as 401K plans lose their intrinsic and projected values. The innate drive to put their heart and souls into their work, that may have been instilled by their family values and training, is now focused on self-preservation, living for today, building a nest-egg, acquiring "stuff," or some combination of all four. If they have lost their jobs, those with self-respect have to retrain new-technology jobs at a period in their lives when they should be the "senior" staff, mentoring apprentices, and sharing a lifetime of professional wisdom with others. Those who may not be able to retrain or who have been devastated by loss of their job, assets and retirement programs can't cope with multiple disasters at one time. May find themselves performing entry-level jobs or turn to public assistance just to exist. At the very least, this group of individuals, who should be our respected teachers and sages of wisdom, are more often bitter, apathetic, and have abandoned any sense of social conscience (because the community abandoned them) for apathy, ambivalence, and acceptance of mediocrity as a way of life.

THE GEN-X WORKERS

Whether we have not provided them with effective mentorship, the media has skewed their value system, they have created their own value system, or they have simply mutated from a combination of all these stimuli. The current generation of trade and craft workers have an entirely dissimilar approach to craftsmanship, excellence, and loyalty than did their fathers at the same ages. I just attended a seminar where the keynote speaker focused on this very subject and presented some astounding statistics that support my conclusions. The first difference is that this generation often leaves the nest and moves to a GU (geographically acceptable) location such as the beach, mountains, big city, etc., or they go where they can make the most money. Nevada led state growth for the nine years and most of the craft and trade workforce has emigrated from other states, because wages were high, jobs were plentiful, and pastime activities were attractive. Similar migrations happen regularly to areas affected by natural disasters like hurricane-ravaged areas of the Gulf Coast. Their fathers probably made geographic moves only to the suburbs or to locations that offered long-term opportunities like California or Texas. Gen-X'ers are prone to pull up stakes and bail when they find a location that is more GU or the abundant work runs out. They typically repeat this cycle many times during their lives. In my observations, many of them live for today because the "loyalty" model is no longer viable, the world is a smaller place, and self-gratification are the predominant motives. They are in continual search for the *Rush*, regardless of where it might take them. As I see in my sons, work is a means to an end.

Those who do have an advanced work ethic are either self-employed or ensconced into the few companies that still value craftsmanship and know how to retain these unique individuals. The rest are doing mediocre work for enough money to support their lifestyles and they have no compunction about moving from one job to the next when employers demand more than mediocre performance and behavior from them. A construction foreman told me, several years ago, that much of their workforce had criminal records and it was okay, because they knew their proclivities and would supervise them accordingly. The workers were virtually "under-guard" all the time, but they were also accustomed to working under guard. Those without criminal records were more unpredictable in their behavior! Does anyone else see this scenario as bizarre and uncivilized? Most of my consulting clients believe that they can be compelled to hire anyone with a pulse to fill their hourly positions because higher expectations will only lead to disappointment. They lament that they have to have warm bodies to do their skilled and semi-skilled tasks and that the employers understand that they will have to continually invoke costly checks-and-balances to ensure that customers do not receive defective or mediocre product.

THE BABY-BOOMER PROFESSIONALS

This be me. My life story is chronicled in the Introduction to this book and is reasonably typical of my generation, up until about age 35. In 1980, I quit Ford Aerospace and did not realize the corporate retirement I was born and bred to do work toward. In retrospect, neither did many of those who stayed loyal, as the company was eventually sold, and their Ford Motor Company entitlements disappeared. In an equally brilliant career move, I joined the petrochemical industry just as it was heading to a new low in its checkered history. I watched as obscene fortunes were made in the 60's and 70's in the oil boom, and then participated in the loss of those fortunes in the companies I chose during the 80's. I hitched my star to two companies that are no longer in business and two that "right sized" and eliminated my positions, as domestic oil exploration busted. Being an optimist, I never had any expectations of collecting a retirement package, so I wasn't very disappointed when I had to play professional hopscotch. I quickly lived the five stages of grief each time I was riffed, and then moved on to the next challenge. I did not realize that my resiliency was not the norm. I always gravitated toward inspired entrepreneurs and was insulated by many of my contemporaries who were bitter at the world for dealing them layoffs and insecurities. As my wife will tell you, the entrepreneur (that I have become) knows that failures are merely opportunities to overcome challenges in our quest to build the better mousetrap. While I am genuinely uncomfortable and anxious during my periods of teetering financial disaster, I know that tenacity and belief in my dreams will turn around those moments of horror. What about those who never learned Colonel Kurtz' lesson that I quoted in Chapter One, "Horror has a face... and you must make a friend of horror?" These Boomers are bitter, resentful, and downright angry that they had a huge mortgage on the house on the cul-de-sac when they got laid off from MCI. Many of them had just gotten their composite golf clubs and were in the process of refitting the bass boat when the gravy train unexpectedly ended. They had built on the college education afforded them by their parents, worked hard, raised their 2.5 children, and were reaping the rewards of their hard work. For many, a bubble burst around age 35 to 50 and their leveraged financial empire was poised for collapse. Our country had reached a convergence of overly-inflated wages, excessive benefits, executive greed, unrealistic expectations, jobs moving offshore, and our credit card balances reaching critical mass. Gated communities of the 1990's were dotted with for-sale signs, credit counseling offices and bill collectors.

What the hell? Our parents were retired and living comfortably in senior's community on their savings and Social Security. They had provided for us and we were now providing for them with our SSI contributions, but we had accumulated debt instead of savings. Our kids were in expensive colleges and driving stylish vehicles that we were paying for. We forgot that our parents' luxuries were very modest compared to what we demanded for ourselves and our families. The pension my parents got from Canada Dry will not be there for my generation and the Social Security benefits they spend modestly probably will not adequate support our life styles, unless we had the sense to start our own retirement savings accounts. In fact, if professional jobs keep moving overseas, there won't be a government insurance program for us at all. If we speculate badly in the stock market instead of investing conservatively in a 401K, we'll have nothing at all when we can no longer earn a living.

Instead of being a generation that evolved one level of comfort above our parents and were on track for rewarding retirements, many of us are swimming in debt and retraining for new job positions, five to ten years before we thought we would be playing golf every day. Were there causes beyond our control? Sure. We never anticipated that corporate greed and corruption would reach the level that lead to the downfall of entire corporations and the end of entitlement programs for many others. We were unaware of the potential downsizing that resulted from massive mergers and acquisitions. We did not see our high-tech jobs moving overseas and competing with H-1 immigrants for existing jobs. We did not anticipate that Communist China would learn how to exploit capitalism and use the best of both economic systems to leverage themselves to take over the world (its coming, sports fans).

Those factors notwithstanding, there was a lot that we Boomers should have controlled, but ignored or abdicated. Credit providers made it seamless for us to get that High-Def TV, All-Terrain Vehicle, and laptop for each kid. When credit card debt became overwhelming, rising home values provided an opportunity for sub-prime mortgage companies to lend us the equity in our homes to pay off the credit cards. Now we have no equity in our homes. We are dedicated to our careers but were shocked when we were replaced with younger and less-expensive workers. When we read the last few chapters of this book, we were stunned at what our kids are watching on TV and the lyrics they are singing along with their smart phones. We are the ones who either demand or allow extreme sports, TV, and reality events. Because our lives are consumed with careers, juggling debt, and self-aggrandizement, we have been oblivious to the creeping mediocrity we have allowed to become part of our lives and our society. If you throw a frog into a pot of boiling water, it will jump out. If you place a frog in a pot of tepid water and gradually raise the heat, it will be scalded to death. We, as a generation, have been in the pot of hot water for years. The heat keeps rising and we haven't the sense to jump out.

GEN-X PROFESSIONALS

If my sweeping assertions in previous chapters gave the impression that I think all Gen-Xers are slackers and have no drive, which simply is not the case, a spin-off of Extreme TV, Donald Trump and a host of get-rich-quick programs, then our current up-and-coming execs are highly motivated and driven. If I implied that they are not stewards of our social ethics, then it is not. We have spawned a generation of liberal protectors of human, animal and environmental rights who border on fanaticism for their respected causes of conscience. If I gave the impression that they aren't diligently pursuing higher education with the motivation to become consummate professionals, it was unintended. College campuses are still grinding out graduates at a record clip. If I implied that global anarchy would reign when the Boomers all pass on, I know that is not the case. Humanity will endure, as it has for millennia. What I will assert is that the motivations for Gen-X to be successful are manifested almost solely in the "I" with very little regard for the "we[20]." Their apparent value system is about self-aggrandizement, living for today, experiencing the *Rush* as often as possible, and "He who dies with the most toys, wins!" Okay, pop-psychology majors, I know that contemporary teachings stress that we must focus on the "I" and not the "we." We cannot be fully functional unless the "I" is physically and emotionally healthy. We cannot be part of the community if the "I" is codependent or lack in our basic needs. I agree. What often gets lost with the current generation is that the "I" must also be fully accountable for our actions before we can be whole. The Boomers did create the welfare state, but most of us did not intend to leave dependency on the government and lack of social accountability as our legacy. Instead of holding our mistakes out as learning moments for improvement, we glorify them as excuses. Corruption is inevitable, so

---

20   Those who are invested in environmental, animal rights and other causes may not be as altruistic as they appear on the surface IMHO.

why not give-in to the "Dark Side." The Boomers did begin the phenomenon of transferring responsibility for our social ills to the government, the enemies of the state, corporate executives, and our parents. In doing so, we have unintentionally lowered the expectations and accountability of our children and have laid the foundation for perfecting mediocrity by sensationalizing corporate and political scandals. Instead of the junk-bond scandals of the late 1980's leading to better fiscal accountability, it set the stage for larger scandals like Enron and WorldCom. Those scandals led to the Sarbanes Oxley Act, which sets new standards for corporate accountability. I will wager my gold teeth that there are currently up-and-coming Gen-X professionals who are perfecting work-arounds for Sarbanes Oxley. Instead of Bill Clinton's affair with Monica leading to shame and moral reform, it redefined sex for Gen-X (and future generations). The lesson from his immoral behavior is that oral sex is not actually sex, according to current high school and college students. The professional Gen-Xers are rationalizing their selfish behavior by the examples we, Boomers, set for them. Again, I say that, without a national crisis like World War II, Americans have a tendency to take the path of least resistance. We are creating an ever-decaying social morality that is playing into the hands of China, India, and anyone else who watches our TV channels to identify our weaknesses. Gen-X is building on the lessons of the Boomers and is skillfully perfecting the avoidance of accountability.

We can't solve our business or societal problems until we stop attempting to perfect mediocrity[21].

---

21   I've intentionally not taken the time to discuss the next generations after Gen X because I can't stomach researching them until we have eliminated the social acceptance of those in the picture above.

Chapter 10 - Successfully Avoiding Personal Accountability

Is this person answering your web posting for a marketing executive?

VOLUME III
THE PATHOLOGICAL BUSINESS MODEL: MANAGEMENT VERSUS LEADERSHIP

Thus far, I have made some pretty sweeping allegations about the ills of society and corporate America. Having worked with more than 600 companies over 40+ years, I have accumulated a wealth of experiential and anecdotal data to support my theses. A colleague offered to me the concept of business as "pathology" and I will make my case for the use of that word (pathology is the series of processes that diseases follow from incubation to conclusion) in this chapter.

As you may have gathered, I believe that history is our greatest teacher. Moreover, I am convinced that we are squandering the knowledge others paid so dearly for, by us ignoring the wealth of cause-and-effect data that is available to us. In fact, I will devote the entire next chapter to this topic. In order for us to understand our current business pathology, I am going to take us on another quick historical journey, this time from the perspective of organizational infrastructure and its more recent evolution.

From the earliest days of our colonization until the last days of the nineteenth century, America followed the European traditions of goods and services being produced by individual craftsmen, in their own shops in their own communities. There were large organizations, like Colt and Winchester, which produced significant quantities of goods, but they were the exception prior to the Industrial Revolution. Samuel Colt's gun factory in Hartford, CT, started as a sole-proprietorship and grew into a large organization because of the demand for the revolver by warring factions. Winchester also began modestly until the Model 66 rifle became known as the "gun that won the west." The company had to find a way to keep up with the demand for the rifle and the ammunition to support it. It would be accurate to state that many of our better-known industrial giants started as entrepreneurships or family businesses and their growth into sizeable corporations was a result of demand for their products far exceeding what a mom-and-pop operation could produce. Within this business model, there is a longevity issue that exists even today. The charismatic geniuses that started these companies could not be cloned. Thus, many businesses failed (and continue to fail) because the transition to corporate or "offspring" ownership that often cannot reproduce the "magic" the founders brought to the success of their products or services.

Other entrepreneurs of the early industrial revolution began their businesses based on advances in technology, rather than mass-production of staples of civilization, like guns. Examples include textile manufacturing and farm machinery. There were few textile mills that evolved from granny's attic and even fewer tractor companies transitioned from buggy-whip craftsmen. While Colt and Winchester could build on the mystique of their histories, the new-tech companies had no such heritage and had to develop their own identities. As an interesting aside, I was reading the history of Winchester and found an intriguing piece of data to support the importance of brand recognition for the older companies. According to their web site, the Winchester brand had been bought and absorbed by Olin. In the 1980's, Olin realized that they had lost their identity with sports enthusiasts. They created a new enterprise to capitalize their heritage, which is again known as Winchester! A similar story may also be true of Ford, since there is often another "Ford" running the company and attempting to salvage what is left of Henry's dream. Curious food for thought!

Since we became aware of the history of sweat shops in New York City's garment district, they might be a good example of how the American industrial corporate identity began. Mass-production, as it turns out, can be relentlessly repetitive, boring, dangerous, and unfulfilling work. Pieceworkers were paid for reproducing the same article of clothing, over and over, with an ever-increasing demand for higher and higher output in less and less time. The combination of speed, tedium and receiving the most meager of compensation was a formula for poor quality, accidents, and high turnover. None of these was an issue for the entrepreneur because they could discard the rejects and Ellis Island was supplying a never-ending stream of new workers (Even if you were a journeyman machinist in the old country, you may have had to work as a laborer in a textile mill, while you were becoming settled and acclimated to a new country). Under the "Management by Fear and Spend no Overhead" model, there was no incentive for any process improvements for safety or for employee retention. Underperforming workers, or those who got injured on the job, were easily replaced with fresh meat from the long lines outside their employment offices. Whether we hypothesize that this management style evolved from feudalism or Elizabethan morality, it certainly grew and prospered in the early days of mass-production in America.

When the workers had enough of being treated inhumanely, labor unions formed to take up the plight of the masses by organizing and establishing more sane work rules. While union leaders would have been beheaded in feudal times, this was now twentieth-century America. They forced a paradigm shift to begin at the senior corporate level. Feudal bosses could deal with employee turnover, but not with work stoppage by their entire workforce. It would have been wonderful if the sweat shop operators had seen the union movement as an opportunity to reform their methods and reform their early-Ebenezer Scrooge models. Unfortunately, unions were highly resented because they limited the amount of tyranny employers could employ to enrich productivity and profitability. With the proliferation of unions and collective bargaining agreements, a clear adversarial relationship was sweeping the nation between management and workers. Battle lines were being drawn where management would concoct schemes to avoid or evade union demands, while the union leaders invented more and more demands to offset those they could not enforce. This war raged for over sixty years until unions became so powerful and corporations became so adversarial that we priced ourselves out of the market as producers of affordable goods. Just as the Chinese are doing currently, the Japanese learned to watch us fight, reverse-engineer our products, and built them at a fraction of the price that we could in a union shop (If you take nothing else from this book, remember, history ALWAYS repeats itself).

When my dad was selling soda, his union grew so powerful that, while other companies were doing away with commissioned sales people and using truck drivers to sell products, he and his cronies were being paid unreasonably high wages, with excessive benefit programs and unconscionable amounts of paid time off. He had an iron-clad pension program and, with his seniority, he was working only about 6 hours a day. On a crisp spring morning, when he was 58 and too young to draw any union benefits, the company closed its doors because it could not compete by paying commissioned sales people to do the same job that truck drivers did in other soft-drink organizations. He spent the next seven years (until retirement age) as a very bitter person, working odd-jobs to make a living. My mother had to go to work so they would have medical benefits and a steady income. If it wasn't for his pension that actually survived, I might have become part of what is now dubbed "The Sandwich Generation," who supports their parents whose pensions did not survive corporate and union greed and pension-fund-mismanagement.

Before our sweat shops moved to Central America and Asia, there was little incentive to modify the adversarial management style we had honed so skillfully. Fortunately, over the last few decades, many workers who had been riffed from their jobs or displaced by product manufacturing moved overseas became our new entrepreneurs. These individuals had fresh scars that had yet healed. Their new rally-cry for corporate leadership was to "not repeat the mistakes of their former employers." Some of these would become the "ex" companies (such as in NYNEX, which was a regional phone company formed after the breakup of the Bell dynasty), meaning ex-employees of corporations that had closed, imploded or priced themselves out of their market. I've worked with many "ex" companies over my career. The principals are always highly motivated to learn from the mistakes of their former employers and to build a company that will never have the kind of callous corporate bureaucracy they had experienced in the past. Unfortunately, these well-intentioned entrepreneurs were not trained in avoiding the mistakes that create impersonal and inflexible corporate infrastructures. As they were preoccupied with cash flow, acquiring a customer base, and bringing their product to market, they began adding "employees" who have no idea that their new bosses intend was not from the same ilk as their previous employers. The fatal mistake that most entrepreneurs make is that they seldom take the time to capture the vision of how they want their companies to run and communicate that shared vision clearly to everyone who joins the firm. They do not know how to codify their values and create behavioral parameters that ensure everyone is treated fairly and with dignity. Having spent a great deal of time at Dell Computer, one of my seminal observations is that, since Michael Dell went directly from his college dorm to running his company, he was never jaded by corporate America. As a result, he had the innate sense to document his vision, mission, values, and rules of conduct. As large as the corporation is, today, if you survive the interview process and become an employee of Dell, you will be trained in the shared vision, mission, values, and rules of the road before your probationary period ends. If you enthusiastically subscribe to each of them, your potential for growth is endless. Unfortunately, we now have Dell Computers on the shelf at Best Buy and Walmart. Michael took a hiatus from his leadership role and his heirs forgot the Dell Direct model. He's had to spend his personal fortune to buy his company back, but the damage is done. Dell is no longer #1 in every category in Consumer Reports. It is another "commodity" computer with a customer service call center somewhere in India. Not only are they not the brand leader any more, but their customer service (IMHO) has also become among the least responsive. I have recently purchased a Dell laser printer, a dell desktop computer, and a Dell tablet. Two of the three have caused me to spend hours on the telephone and email with someone named Todd or Chuck, whose accents are so thick as to be nearly incoherent. Most of the conversations with Todd and Chuck are spent with going through their call center script and avoiding escalating my call to someone with some cognitive reasoning. What is most infuriating for me is that, around the turn of the Century, my team was working at Dell and their call center in Round Rock was the benchmark of customer service! Remember the rise and fall of the Roman Empire?

Meanwhile, at company "ex," the new employees are creating their own informal organization with its own set of values and its own hierarchy. This is what they had to do to survive in previous corporate environments, because they always had adversarial relationships with management. Guess what. Because the entrepreneur and his inner circle almost always neglect to create a planned business infrastructure and open communications channels, within a few years they are fighting the same bureaucratic inefficiencies as in their prior companies. They don't communicate anything but product and customer issues to the employees and, without a master plan, employees do their best to design pragmatic solutions to overcome the problems-du-jour. Pretty soon, tribal knowledge runs the company, silos are built to avoid becoming involved with problems of the other departments and moats are dug between the workers and mahogany row. Now for a sad statistic: Of the 600+ companies I have worked with, in my opinion, only two have had the vision and tenacity to truly avoid becoming a bastion of traditional management-versus-worker infrastructure. Hence, business pathology: Doing the same thing over and over, and expecting different results is the definition of insanity.

Beyond the unintended management by abdication that I just described, there are several new iterations of corporate management-by-greed. In these corporations, there absolutely is a master plan, but it exists to extract every ounce of profit from each employee, while painting a façade of being innovative and progressive. The program sold to the employees is either a carefully crafted hermaphrodite of a fiscal plan for the principles to accumulate massive cash-flow and then bail out or it is the result of reading one of the many books available about the successes of the most "powerful" (meaning ruthless) business moguls. It is my contention that an inescapable result of our extreme, vicarious, and *Rush* –driven society is to accumulate personal wealth with little regard for how it impacts those who you employ to build your value proposition.

The dot.com disaster of the 2000's makes my point for me rather dramatically. Ruthless investors partnered with naïve entrepreneurs to produce businesses that had absolutely no intrinsic value. With a proliferation of "ex" employees with clever entrepreneurial ideas, a new model was hatched based on rapid deployment of the would-be product or service utilizing some innovative use of the Internet. A new business model evolved and was sold as the entrepreneur's dream for high-velocity launch, cash flow, and personal wealth. It involved finding "angels" who had plenty of money to launch the next Bill Gates and was intrigued by high-tech innovation. Business advisors created (unholy) alliances with eager entrepreneurs and well-polished business plans became as common as spring flowers. The plans appeared to rapidly build the entrepreneur's breakthrough Internet portal, using startup money from the angels (Portal was the term concocted to describe a virtual store from which consumers could purchase and endless variety of products or services from the comfort of their homes). While the founders were paying themselves salaries to develop and launch the proposed technology, the advisors were out raising successive rounds of venture capital to fund the launch of the portal. In the meantime, the shares of stock were being diluted among the investors to the point that the unsuspecting entrepreneur often lost controlling interest in his own business. Those dotcom's that actually launched a viable portal accumulated revenue as quickly as possible and created massive cash-flow engines that had no intrinsic value. Described as click-and-mortar, they usually had no assets, other than computers in leased office space. Their inventories were often floated in cyberspace, rather than in a warehouse.

Inevitably, the entrepreneur would be called into a meeting by the "board" where they were presented the realities of their folly. Best case, the advisors were poised to announce an Initial IPO to take the company public and sell stock. This scenario established a plan for the investors and owners to become enormously wealthy and retire before those who bought the stock discovered that there was no substance to the company and their investments were in vapor-ware. The more common outcome of this board meeting was to let the principals know that there was no more venture capital, the business would be written off as a loss, and the doors closed. Since these clever attorneys and MBA's had duped the entrepreneur out of controlling interest in the venture, the dot.com victims often found themselves penniless and looking for a job, virtually overnight. I have been told by these "advisors" that their business model was based on the reality that one out of ten dot.com startups would succeed and that windfall would pay for the losses of the other nine. Since the entire house-of-cards was based only on revenue and positive cash flow, there were never any assets (that weren't leveraged) to hurt them and the only real victims were the unsuspecting business person and their supply chain. To complete this scurrilous business model, the suppliers never got paid from the cash-in-motion and they too joined the long line at bankruptcy court. Oh, yes. I must point out that the Facebook stock hit the market in late 2013 at around $47.00 a share and the company has never made a profit.

The failure of the dot.com model has not dissuaded financial geniuses from inventing newer and cleverer get-rich-quick schemes. One of the most popular approaches is by attempting to imitate the model provided by moguls such as Donald Trump. Whether he is either as wealthy or as ruthless as he is portrayed in the media, the model derived by would-be-tycoons centers around accumulating obscene wealth via short-term deals that can be predominantly cash flow mirages or vapor-ware. Another model is based on mergers and acquisitions that are carefully crafted to rape the assets of the acquired company, eliminate huge chunks of duplicate management, and dishonor the employee base by reneging on benefits promised by the acquired company. The third scenario is probably the most heinous. These players are the senior execs of the Enron's and WorldCom's who (allegedly) actually conspire to commit white-collar crimes to defraud their investors and employees. The final outcome of the most egregious of those disasters is still unfolding for all of us to witness. Will corporate America learn to reinvent our social morality, accountability, and stewardship of industry or will we be motivated to find new and more-clever schemes to rape our workers and defraud our investors? Will business leaders start listening to me and my associates who teach shared vision, shared rewards, personal accountability, team accountability, corporate honesty, and win-win business models, at a ratio greater than two out of 500? Or will they continue to be bought out by foreign investors who have no investment in our future besides cash flow and technology transfer? Can we cure business pathology, as it exists today and implement business models that are so lean and productive that we will once-again become the global leader in technology, innovation, and share the wealth via hard work and accountability? In my humble opinion, we have priced ourselves out of the market for manufacturing goods that are labor-intensive. Our only apparent hope is to seize the energies currently being channeled into extreme behavior and create the most unique workforce in the world, of self-managed work teams that are accountable to each other for corporate and personal success. For those of you who lead La Vita Vicarious, you will not succeed by reading about Michael Dell, Jack Welch, and Donald Trump, and then try to replicate their success. Their genius and dedication was complimented by timing, context, environment, their unique networks. and random good fortune that you cannot duplicate by skillful design and intent. What you can learn from them is that each amassed their personal fortune by creating jobs, bricks-and-mortar, and value, not vapor-ware.

There was an article in Entrepreneur Magazine, October 15, 2012 entitled *Jack Welch on How to Manage Employees*. Since the Jack Welch Management Institute is considered one of the premier training grounds for would-be business leaders, you would think they would know that you lead people and manage processes. Also, if you are building your business model on Jack's, remember his genius in turning around General Electric included buying a credit card company to finance Six Sigma[22] and his other signature "success programs."

---

22   Six Sigma is touted as a set of techniques and tools for process improvement. It was developed by Motorola in 1986 as a statistical modeling tool for large scale process improvement. Today it is a very lucrative business for consultants and trainers who are misapplying it to businesses that haven't the volume to utilize it.

VOLUME IV
THE KAIROS MOMENT: DO WE REALLY LEARN FROM HISTORY?

"Wisdom is the right use of knowledge."
– C.H. Spurgeon, 19th Century Theologian

Kairos is a Greek word that means the suitable or the right time at which something should be done. The Kairos Moment is defined in theology as a single fleeting chance for the enlightened to become students of profound knowledge created by catastrophic events. In these moments of focus, students of The Kairos Moment have extracted great knowledge about cause, effect, recovery, and human behavior. Scholars have produced some of their most profound writings from a Kairos Moment. Their writings have been used to convey great comparative insight and wisdom to those who seek knowledge from history.

I have trademarked The Kairos Moment as it applies to business and have expanded on the theological definition to include summons for contemporary business leaders to convert knowledge obtained from external tragedies and from their own disasters into the wisdom to implement insightful and beneficial change.

Natural and manmade disasters are unplanned events that cause ordinary people to perform extraordinary tasks. From events such as 9-11 and Hurricane Katrina, we have heard recounts of numerous heroic deeds accomplished by humans who responded to the urgency of the moment with superhuman acts and resolve. While these stories often cause us to well-up with emotion, if we do not extract the key lessons from these events, the greater benefit of their courage and sacrifice is lost in the parable.

Disasters also often create a moment that triggers the transformation of knowledge into wisdom for solving <u>impossible</u> problems. I remember an event in our town where a rather frail housewife lifted a car from her husband's leg when a bumper jack collapsed and pinned him. Solving these types of "impossible" problems may be attributed to an enormous adrenaline boost triggered by the shock of the moment. In the next chapter, however, we learn lessons from the near-disaster of Apollo 13 and how the ground crew overcame the impossible task of designing a CO_2 filtration system from the materials available in the Apollo Command Module and in the Lunar Lander, to keep the astronauts alive as they flew their crippled spacecraft home from the moon. The rest of this book will make the case that the story of Apollo 13 is more than a great movie, it is rich in business lessons that can help us all avoid more mundane catastrophes in our own universe.

The first Kairos Moment lesson we will learn from Apollo 13 is ***situational awareness***. By studying the events of the disaster and ensuing dramatic recovery, it will become clear how the tactical disaster recovery team in the Mission Control Center were able to quickly create adept robust solutions to problems that had never before been anticipated or simulated.

What crisis or disaster has befallen you that could have been a moment of clarity and ***situational awareness***? Are you conscious of having experiencing a Kairos Moment or did the tragedy cloud it for you?

In the late 1990's, The Federal Reserve Bank of New York implemented a systematic program of continual process improvement with the purpose of elevating customer satisfaction, reducing cost, and minimizing operational risk. They chose to implement the International Quality Standard ISO 9000 as the underpinnings of this new initiation. ISO 9000 provides a proven framework of documenting business processes, working to the documented procedures, regularly assessing them for effectiveness and continually improving them. The examples of the techniques needed to implement this proactive methodology in a financial services environment were only available in Europe and Asia. Converting tools previously implemented mostly in manufacturing environments in the USA would be a formidable task for the Fed. Their relentless pursuit brought them to successful registration to ISO 9000 and they are still one of the few banking institutions in North America with this prestigious certification.

Their Kairos Moment came after the catastrophe of 9-11-2001. Banking, as they knew it, had to be reinvented to deal with the disaster. Just as Apollo 13, they had to invent and implement robust processes that had never been modeled or simulated. They did it at the moment that many other financial institutes were paralyzed and unable to conduct business.

The Kairos Moment is a stunning event or circumstance that compels the enlightened to engage in deep introspection and re-examine his or her state of affairs. The dramatic event can present to the individual a rare and life-altering opportunity. He or she can seize The Kairos Moment and use it as a forceful catalyst to come out the other side: weathered, wiser, and stronger. Those who do not grasp the Kairos moment often shut down and become victims, totally at the effect of the disaster. Still others become looters. The choice for those who will embrace the Kairos Moment is clear, compelling an opportunity to grow to new heights of personal and professional excellence.

The lesson from the Kairos Moment for the Federal Reserve Bank of New York was **enlightenment.** The actual quote from Quality Director, Carol Nowicki, was, "If it wasn't for the discipline we learned from ISO 9000, we might not have survived 9-11." Rather than reacting to the tragedy emotionally or trying to invent new procedures, they called on their training from the ISO 9000 implementation. They assessed the needs, drafted interim operational procedures, and implemented them. Rather than abandoning their training in a moment of panic and reactive behavior, they followed the ISO 9000 implementation model and built robust procedures, based on proven methods. They had reached a state of **enlightenment** unknown to them before 9-11 to create the environment that would turn tragedy into successful recovery.

What crisis or disaster has befallen you that could have been a moment of clarity that turned *situational awareness* into **enlightenment**? If you did become aware, did you experience **enlightenment** as a result of the Kairos Moment?

In January of 1986, the world experienced another wake-up-call in the form of the explosion that destroyed the Space Shuttle Challenger. We watched in awe and horror as a leaky seal in the solid rocket boosters began a chain reaction that sent the spacecraft and crew to the bottom of the Atlantic Ocean.

In this instance, the entire leadership of NASA experienced a Kairos Moment. The effect of the tragedy touched all levels and put the Space Shuttle Program on hold.

The findings of the Presidential Commission on the Challenger disaster revealed a series of process and human failures that led to the accident. Among the findings were:

- Quality management processes were circumvented to ensure launch schedules were met
- The process for making a launch decision was flawed and safety could be compromised for expediency
- Astronauts had been removed from the management structure of NASA. Instead, bureaucrats who focused on budget and schedule made life and death decisions[23].

A visualization of the Kairos Moment. This photo was taken at the author's home on the afternoon of 9-11-2001

---
23 Recently, I have spent some premium time with two young(er) Space Shuttle Astronauts. To my absolute horror, none of their training has included any of the values, processes and core tenets of Project Apollo. They both suggested I propose teaching the Apollo Business Model to the current NASA bureaucrats!

The result of the findings was *action.* The process for installing seals in the solid rocket boosters was revamped. A new safety organization was commissioned. The process for committing to launch was overhauled. Astronauts were once again placed in key management roles. The disaster of the Challenger has led to *action* that has resulted in the flawless launch record since 1986.

What crisis or disaster has befallen you that could have been a moment of clarity that turned *situational awareness* into *enlightenment* and enlightenment into *action*? If you did find *enlightenment* from the Kairos Moment, did you have the courage to take *action*? Were you prepared to turn *situational awareness* and *enlightenment* into *action* to ensure that the moment of clarity is not lost and the expensive lessons of the tragedy are turned into tools of problem avoidance and process improvement? Learning how to take appropriate and timely *action* will enable your organization to turn its disaster lessons into tools of continual improvement, as NASA did after the Challenger accident.

The Kairos Moment presents a compelling case for utilizing known success tools, proven by unconventional means, to model highly effective learning strategies. How could the NASA team on earth so many miles away, help the Apollo 13 crew solve their myriad of life-threatening problems? It was accomplished through this combination: simulating precisely in Mission Control the conditions the crew confronted, utilizing intensive communication and brainstorming, and relying on learning models developed in non-crisis conditions. While the crew of Mission Control may appear larger than life on celluloid, I was there in Mission Control on the first night of the Apollo 13 disaster and I experienced a Kairos Moment that has formed the value system that I live to this day. You may have guessed its foundation by now. That is, always look to history to avoid repeating mistakes. What we have yet to discuss is how to apply the wisdom of history to form creative solutions to impossible problems.

Mission Control Houston after the successful recovery of Apollo 13

How could today's business leaders utilize such a disciplined approach? Are they? If not, why not? When terrorists attacked on 9-11, each of us, perhaps without realizing it, went through The Five Stages of Grief[24]: denial, anger, bargaining, depression, and finally, acceptance. But what then? Did we seize the Kairos Moment to optimize our situation? If not, could we learn a step-by-step process for doing so? Instead of just surviving a disaster, how can The Five Stages of Grief be turned into a Kairos Moment? With enlightenment, we can learn how to turn the last three stages into *situational awareness, enlightenment, and action.*

Traditional business school teachings are based replicating success models. It is my fervent belief that knowledge experienced in disaster is overwhelmingly more practical and applicable than abstract knowledge derived from high-profile success stories. Disaster is truly a better teacher than business models based on the success of high-profile business leaders. Creating the unique chemistry of leadership, opportunity, and talent cannot be replicated from a textbook formula. Disaster, on the other hand, can provide *survival lessons,* based on moments of terror that we all have experienced, reaffirming the significance of a clear vision, effective operational processes, and ongoing training.

Catastrophes provide expensive lessons of survival that have lasting and profound impact on those touched by the events. If these events are so profoundly educational, why then do we not systematically transform that knowledge into sound business wisdom?

As an exercise, perhaps we should treat the music lyrics and television show plots in Chapters 7 and 8 as disasters and look within them for valuable lessons from which we can turn-around our journey to perfecting mediocrity.

---

24 A behavior pattern chronicled by Elizabeth Kubler-Ross in her breakthrough book *On Death and Dying.*

VOLUME V
THE HUMAN LESSONS FROM PROJECT APOLLO

> *"We can't solve problems by using the same kind of thinking we used when we created them."*
> Albert Einstein

At, the risk of repeating myself too often, for many years I was relatively unaware of the personal and professional impact of my involvement and limited contribution to that moment in history of Apollo 13. Even though the questions "*What was it like to be in Mission Control during Apollo13?*" were frequently asked by business associates and friends, they were answered with technical data, statistics, and process details. There may have been some ego woven into the answers (and, certainly, the stories have become embellished over time), but I was mostly conveying that **we just did what we were trained to do.** After the adrenaline rush of the first hour of monitoring air-to-ground and flight controller conversations, the realities of my training became obvious when I went to Sub-stores and dutifully filled out a requisition form for Polaroid film. Although the problems were immediate and the situation was grave **we all just did what we were trained to do**. Why else would I have taken the time to fill out a requisition form in the midst of a crisis? The answer is that **effective process procedures and training prevent anarchy and wasted effort.**

We found tactical solutions to the problems at hand and implemented them the same way we found a strategic solution to the challenge that President John F. Kennedy said, "…this nation should commit itself to achieving the goal, before this decade is out, of landing a man on the moon and returning him safely to the earth." Or the very best I heard stated, "For centuries man has dreamed of traveling to the moon. We just made our minds up and did it."

CHAPTER 11
APPLYING THE LESSONS LEARNED

In the ensuing ten years after Apollo 13, I was fortunate to have been assigned to perform capability audits for organizations such Digital Equipment, Data General, Ampex, Hewlett-Packard, and Sangamo. I also conducted audits at companies that had been concocted to take advantage of NASA's policy of giving preferential treatment to minority business enterprises. Many were viable businesses, but some were obvious scams perpetrated to exploit business minority people and unfairly win government contracts.

As prescribed by our contract with NASA, I used the same quality audit checklists for all companies. Since quality requirements and checklist contents were in the public domain, almost all companies were prepared for my visit and had choreographed a tour of documents and procedures as evidence of compliance. Each exhibit provided would correlate to a check box on my forms. At the end of a typical audit, most of the requirements had been marked as compliant and automatic approval. After a pleasant lunch, I was on my way back to the airport with another notch in my gun for approving one more "happy supplier."

Constantly influenced by Apollo 13, my left-brain was challenging my right brain and auditing suppliers on the checklist started to really make me uneasy. It was the same disconcerting feeling I got when a sales person described the fabulous warranty and extensive service department they had. Why do they need a fabulous warranty and extensive service department if the product was inherently reliable? I couldn't quantify what was causing the discomfort until I realized that a compliant checklist at Ampex and at Bubba's Electronics Company were significantly different in meaning. It was apparent when touring Ampex; they had a quality manual and personnel following written procedures. So it was at Bubba's.

The subtle difference was that at Ampex, the people were obviously just *doing what they were trained to do*. At Bubba's, the records were completed to indicate compliance, but the workers appeared to be operating by tribal knowledge, not written procedures. What they said they did and what I observed them doing were two different scenarios. In fact, in some of the more difficult audits, I was not allowed to talk to the workers, only to the quality manager or sales person. Those conflicted environments always caused my caution and warning indicator to glow red, but I had little choice but to follow the checklist.

One fine Saturday morning, I was on my way home to Houston from San Francisco on a brand new DC-10 jet. It was one of National Airline's inaugural flights with a new wide-body aircraft. My seat was spacious and comfortable. The ambient noise was almost non-existent, compared to the DC-9 I had flown out on. Even the meal service was extremely edible, for a change. Both aircrafts got me to my destination in about the same time, but one was of significantly greater in value and the prices were the same.

I had spent the previous week auditing companies that were proposing to fabricate custom printed circuit boards for us. Fabricating a circuit board requires chemically etching copper away with powerful acids from a "sandwich" of copper with laminated fiberglass in the middle. It also involves plating metals such as nickel, tin-lead, and gold on the exposed copper. The entire process is very nasty. All of the companies I had visited met the requirements of the checklist.

Somewhere over West Texas and after a hot chicken sandwich, I had an epiphany similar to that of a Kairos Moment; I didn't know what to call it when it happened. For the first time, however, I had become aware of the difference between compliance on a checklist versus meeting the needs of the customer. For some reason, reviewing the checklists from the previous week at 39,000 feet evoked a moment of great clarity. I had two virtually identical checklists in front of me, yet the companies they represented were as different as a DC-9 and a DC-10.

Let's call two of the companies I visited Alpha and Bravo. Alpha's facility was typical of most circuit board shops. The air was filled with the choking smell of acid. The wooden floors were wet with spillage. The employees wore dirty smocks with holes burned through by acids. The quality manager at Alpha quizzed me on how closely we inspected the tolerances for under-etching (leaving too much copper), over-etching (removing too much copper) and plating thickness (the amount of nickel, tin-lead, or gold added). This line of questioning made me wary that they didn't always produce products that met our specifications. He hurried me past barrels heaped with scrap boards and a locked cage with discrepant products awaiting approval of quality concessions by their customers.

Bravo's plant had no objectionable odors in the air. The plating and etching tanks were in containments that minimized spillage. The employees wore clean blue smocks, safety glasses, and gloves. There was no evidence of deterioration in any of their clothing. Bravo's quality manager asked only technical questions and if we preferred the circuit boards individually wrapped or in stacks of ten. His questions left me with a sense that they wanted to meet our stated and implied needs. There was a noticeable absence of barrels with scrap material or a withhold cage with discrepant products.

Under the microscopes in the two respective QA labs, the etching and plating of the sample circuit boards offered for my inspection appeared to be identical. In the Alpha QA lab, there was an old and rattling window air conditioner that made it difficult for us to carry on a conversation. I did not sit down because of the discoloration on the chair seats. The QA Lab at Bravo was brightly lit and environmentally controlled to ensure stable temperature and humidity for conducting precision measurements. The quality manager at Bravo informed me the controls were necessary because the fiberglass substrate can absorb moisture, causing inconsistencies in measurements. This gem of information would be transmitted to the design engineers on my return to Houston because parts of our warehouse were not sheltered from the terrible Gulf Coast humidity. The furniture was older, but there was no hesitancy about taking a seat.

How could I turn in identical checklists when Bravo was clearly a superior supplier than Alpha? Why was I always so at ease when working at Hewlett Packard or Ampex but so uncomfortable at Alpha and Bubba's? The light of wisdom had begun to illuminate from my quality training. ***There is a significant difference between providing a minimally compliant product or service and providing one that consistently delivers value the customer.***

I was reasonably confident that we could receive acceptable products from Alpha. My evaluation of their processes also made me anxious that we would receive just enough discrepant products discovered at our receiving inspection lab for us to deal regularly with rejected materials and to tighten inspection controls. When suppliers have a history delivering nonconforming commodities, there is a greater possibility that a random defective product might reach the user and lead to a latent failure.

My expectations from Bravo would be to have shorter inspection times at receiving because of the consistent history of zero defective parts. It would also defy all rules of quality assurance to expect a defective part to find its way to the end user. Let's see, would I like to do all of my traveling cramped in a DC-9 with a box lunch or stretched out in a DC-10 with a hot chicken sandwich? I would still get from San Francisco to Houston, but which was the better value for the same price? In the case of the circuit board suppliers, which **provided NASA with the lowest overall cost of ownership and with the least chance of latent failure?**

I spent the remaining time on my way home writing a narrative report explaining my observations and conclusions and gave a "grade" to potential supplier's Alpha and Bravo. On Monday, I put the checklists and narrative reports in my boss's in-basket, but it felt as if I had lobbed a grenade over the wall and was awaiting the explosion. The bang never came. He thought the narrative was a valuable addition and forwarded my reports to the purchasing department. The purchasing manager was also a bit of a visionary and placed Bravo on the interim approval list and set Alpha aside. Weeks later, Alpha called to see if they were going to get a sample order. The purchasing manager explained that they did not come up to our quality standards and would not be sent any sample orders.

Shortly afterwards, the fur flew at NASA. Alpha called our contract monitor and raised the roof. They knew they met the checklist but were not approved as a supplier. How could that be? Our contract monitor didn't have an answer, so my department manager and I were summoned to a captain's mass[25] in the Operations Support Wing (OSW) of Building 30. It took a while to explain the entire story about the DC-9 and DC-10, but the outcome of the potentially confrontational meeting was very positive. NASA will send an engineer along on my audit trips, for the next few months, and observe my narrative and grading system. Within six months, NASA approved the narrative and an objective grading system for ground support hardware suppliers. Alpha never made the cut as an approved supplier.

With that victory under my young belt, I was feeling pretty smug. Fortunately, I became grounded into reality when my leaders wanted me

---

25 Captain's mass is a military metaphor for an inquisition held to determine if misconduct has occurred.

to explain to the world what criteria we should publish to objectively evaluate the total cost of ownership of products and services. In other words, what indicators could we identify that would objectively point out the perils of Alpha and Bubba and the merits of Ampex and Bravo? What attributes had I observed at Ampex and Bravo that would predict that they were more reliable and cost effective suppliers?

That challenge was like trying to explain why it felt better to ride in a new Cadillac than in a ten-year-old Pinto or why pizza tastes better than celery. Some serious introspection led me back to the roots of The Kairos Moment, Apollo 13. How did we *just do what we were trained to do* and rescue three astronauts from certain death as routinely and flawlessly as a Swiss watchmaker building his thousandth precision timepiece? What were the lessons of Apollo 13 that gave me the clarity to establish pragmatic rules for evaluating the effectiveness of a company's ability to produce acceptable products and services?

As the rays of light burst through a dark cloud and illuminate a silver lining, lessons from disaster can illuminate a clear path to avoiding future disaster. There were specific lessons to be learned from Project Apollo that, by design or by illumination, caused a business model to evolve that made routine excellence and failure not an option. Apollo 13 dramatically validated our methodologies by demonstrating that trained and motivated individuals are capable of performing heroic and nearly impossible feats when confronted by certain disaster.

If word processors had been available in the seventies, I would have been more productive. Instead, I spent months trying to correlate the lessons of Apollo 13 to my spin-off methods. I trashed reams of handwritten and typed epistles on establishing and evaluating business processes. I erased and redrew many lines and boxes on diagrams of process flows. For a time, I thought my most cogent methodology would be to evaluate the cleanliness of the men's room as an indicator if a company did or did not build quality products (A dirty restroom had virtually always correlated to a substandard vendor evaluation). A clean and orderly restroom had always been an indicator of a superior company. This discovery was statistically accurate, but probably would not be published in a NASA Supplier Auditing Handbook.

Our operational procedures and standards at Philco dealt mainly with establishing process controls to ensure successful outcomes. The manuals were thick with inspection and detection, workers and inspectors, crime and punishment. I wasn't finding the documented evidence to support my observations of what made an organization truly successful. Our tongue-in-cheek rule of thumb is when the weight of the paperwork equaled the weight of the electronics hardware, we are ready to fly a mission. Surely with tons of reference materials at my fingertips, I would find the enlightened teachings that supported man's greatest technological feat, traveling to the moon and returning safely to earth.

My research led to more and more quality control tools and detailed product and process specifications. There were no epiphanies to be experienced in these libraries. The data I was seeking would be found in the events leading to my discovery of the pragmatic methods I had adapted for supplier evaluations. I began mapping my own history and how it had precipitated the enhanced business passion and curiosity within me.

That cognitive introspection led me directly to the root of the driving force. It was (again) the famous speech by President John F. Kennedy that challenged "…this nation should commit itself to achieving the goal, before this decade is out, of landing a man on the moon and returning him safely to the earth." Obviously, the President was talking to me specifically in this address. He was throwing down the gauntlet to get my engineering degree and to be part of making the real vision he dreamed for us as a Nation. I started college in the fall of 1963, so I should have been able to contribute to the lunar landing by 1969. No, wait. The math did not work out. I wouldn't be in the industry until 1968, so I wouldn't have been much help with the early days of Project Apollo. The only logical solution was to leave the family home in Queens, New York and move to Houston. While attending college, I would be able to work at NASA and be a part of the team as early as 1965.

I packed all of my belongings in my Corvair and left for Houston on my 19th birthday. I enrolled in the electrical engineering school at the University of Houston and found employment at Philco a year later. My enthusiasm, not my credentials, got me an entry-level electronics technician position. Our group assembled, wired, and tested the consoles and ground support hardware for Mission Control and most of Building 30. If you sit with me through the Apollo 13 movie, I will name the various modules in the consoles of the flight controllers and tell you what the lights and buttons were meant to do. I'll even point out the subtle inaccuracies in the movie set.

Here I was, contributing to President Kennedy's vision. His call to action was so clear, concise, and resonant that it was at the center of every activity I performed in the 14 years I worked at NASA. In fact, we had very few "employees" at Philco. There were a few working there, just to collect a paycheck. Most of my coworkers shared my passion for space exploration. In fact, I never really understood the need for quality control inspections because the workers were highly motivated in mission success that making a poor solder connection or not checking the accuracy of our wiring was unthinkable to our mission-driven workforce.

Wow. That word *mission* kept resonating in my head. Then, it finally occurred to me that the volumes of quality control requirements in our library had almost nothing to do with the precision workmanship we routinely achieved. Our reliable work output was a direct result of the ***vision*** provided s by President Kennedy. Our engineers designed 99.95% reliability. Statistically, that meant NASA would expect to have 5 failures in every 10,000 missions. For pioneering technology, that's an acceptable risk factor. JFK's ***vision*** was certainly integral to the engineers' tireless hours for designing mission success. We also employed double and triple redundancy when flight safety was at risk (Remember? The Third Floor MOCR was on *hot standby* during Apollo 13.). Along with the design engineers, the technicians, installers, and support personnel also shared the ***vision*** in conducting their daily jobs. In fact, the closest function we had to a *warranty department* was the quality engineers manning their posts during a mission. As I reported earlier, we seldom had anything to do, making us prototypes for the classical lonely *Maytag Repairman*.

Finally, the lessons of Apollo 13 were becoming obvious. A clear, concise, and compelling ***vision*** was the foundation for producing highly reliable products. All the rules and policemen in the world could not accomplish what self-motivated individuals working toward a single common goal were capable of doing. Furthermore, these same people, operating in crisis mode, could accomplish nearly impossible feats in record times. Motivational training or monetary incentives are meaningless to self-directed individuals in a Kairos Moment.

During Apollo 13, JFK's ***vision*** was reinforced with an imperative from head flight controller Gene Kranz. His stated directive to us (The Apollo 13 team) was "***Failure is not an option.***" The two imperatives were etched into every fiber of our beings and drove every action. Our dedication to delivering a compliant product, on time, within schedule, and with no defects was not a contrivance of some clever management scheme. It was intrinsic, genuine, unassailable, and not for sale at any price. During the days immediately following the explosion aboard Apollo 13, many key individuals did not sleep, did not receive any extra pay, and would not go home a minute before their portion of the rescue was complete. Can a compelling vision be this powerful? It's taken me more than 30 years to validate, but it is powerful and *awesome* in the right environment.

Lesson #1 from my first Kairos Moment is that a clear, concise, and compelling vision, that is championed by a true leader and owned by everyone in the organization, is the cornerstone of business success. Organizations that have visions, value statements, and quality policies, developed by the marketing department, that only appear on the wall in the lobby and in the annual report are missing one of the most powerful success tools available to them. Perhaps business leaders aren't able to develop a vision as powerful as JFK's, but Gene Kranz was a senior manager who was not known for his eloquent prose. In his own way, he made "Failure is not an option." a driving force that we could live without coercion or monetary incentives.

Once the cornerstone was laid, the rest of the lessons began to fall into place like the last few pieces of a jigsaw puzzle. Once I understood the motivation of the Apollo 13 rescue team, it was obvious that our successful rescue was, in fact, a direct result of our training. After identifying the result of the disaster, we immediately looked to our documented procedures, flight plans, reference documents, and checklists for guidance. Rather than abandoning everything we were trained to do, we followed the sequence of the documented procedures and modified them to deal with the new scenarios. We did not look for blame. We did not file suit against the manufacturer of the defective oxygen tank. We did not waste time looking for the root cause of the problem. We would do that after the crew was safely home. We didn't panic and try to launch a rescue mission that might have killed even more people because pre-launch protocol and safety rules would have been circumvented.

After the initial shock of the disaster, cool heads prevailed and looked to successful procedures for guidance under new circumstances. It reminded me of the various training videos I have seen on if your auto is suddenly submerged in a lake or river. Will I stay calm and follow the training video, allowing pressure to equalize before opening a window and swimming out? Or, will I panic and try in vain to open the door against the pressure of the incoming water? As police and emergency service personnel will tell you, the only way to survive is to follow established procedures in times of disaster. Panic or reactive behavior will only make the disaster more destructive.

How does lesson #2 translate to other organizations that aren't flying to the moon? It clearly makes the case that companies must have effective process procedures in place to give guidance for their daily operations. Training must be at a level that allows following documented procedures key to accomplishing the vision. The procedures must be *alive*. That is, they must reflect current and changing conditions and be updated as a result of continual improvement and changing requirements. Those who follow the procedures must be trained and continually drilled on their content and importance. Airline personnel are continually in training and drilled in simulators, even though they fly several days a week. Their written procedures are so dynamic and effective that the crews cannot lapse to reactive behavior, especially in a time of crisis. Nearly every potential anomaly in passenger airline travel has been documented, simulated, and trained repeatedly. As a result, the airline industry enjoys the best safety record per passenger mile traveled.

Lesson #3 is that no vision or documented system of process procedures can be effective unless the environment exists to encourage *entrepreneurship*. Yes, entrepreneurship. I've chosen that word carefully, even though it might appear to be paradoxical, especially in a seemingly bureaucratic environment such as NASA.

An entrepreneurial environment is that of self-motivated individuals working towards a common goal. Not only does the leadership provide a clear and shared vision and the framework of robust and documented processes, but there is also a system of risks and rewards that is clear and ever present. During Project Apollo, our risks and rewards were far greater than the monetary connotation usually associated with *risk and reward*. The risk was the untimely death of three astronauts. The reward was winning the space race and getting to the moon first. That set of risks and rewards is hard to match in the business world, but it begins to dismantle the common theory that annual salary increases are a reward and the threat of losing one's job is a risk. During my tenure at NASA, I was probably earning half of what I might have in a private industry. My only fear of losing my job was not being a part of the manned space program, not the family, social or financial consequences.

Encouraging entrepreneurial behavior at all levels is an alien concept to most business leaders. Framed in the context of Project Apollo, my definition might be more palatable. An entrepreneurial employee is anyone who has a clear, concise, and compelling mission that is all consuming. The mission must be so engrained that there is never a consideration for time clocks, overtime pay, sports, hobbies, or distractions when working on the mission. I'm not suggesting work-aholic-ism or dysfunctional home life, I am suggesting that, when at work, an entrepreneurial employee is preoccupied with carrying out the shared vision, not with power, self-aggrandizement, or office politics. The second ingredient is a clear path to follow. There must be a roadmap (process procedures, standards, quality plans, reference documents, and performance data) that keeps the highly motivated and focused individual traveling down a straight path to the desired destinations. There must be clear boundaries of where the ditches and hazards lie on either side of the road. The successful individuals will know exactly where their boundaries exist. They will accept the accountability for getting to the destination. They will learn to ask forgiveness instead of permission when they make entrepreneurial decisions that fall within the boundaries agreed upon. Finally, an entrepreneurial employee is fully aware of the rewards for achieving success and the risks of falling short or failing. There are no such things as annual raises. Compensation is based on real accomplishments. They are responsible for themselves and coworkers for both individual and project success. They are also acutely aware of the consequences of substandard performance. No coworkers are on a free ride or unaccountable for their own actions, thus all boats rise on the high tide.

One clear example of this behavior became clear to me on one of my consulting assignments at Dell Computer. I went looking for one of my clients for a scheduled meeting. His assistant told me he was running late, but to wait. When he came running in the building, he was out of breath and perspiring from the oppressive heat and humidity of August in Austin. He explained that he had just gone down the street and leased some office space for his group. I inquired if it wasn't the job of their facilities group[26]. His response was that he needed the space that week - it was within his budget. He would take the accountability of his actions with his boss. That was a dramatic example of an employee who had a mission, shared vision, clear procedures, and the accountability to ask forgiveness instead of permission when the situation dictated.

---

26   Being late to a meeting was not tolerated at Dell. It was considered an affront to the others' time. It caused me to coin the phrase "You are either on time or you are rude."

Obviously, extrapolating the *entrepreneurial employee* from the spirit-de-corps and work ethic of the Apollo 13 team required the next twenty years of experience and development to present to you as valid. Also, you may have noticed that I have avoided using the words *team* or *teamwork*. The sports team metaphor doesn't work for me because professional sports personnel are highly compensated specialists working for autocratic leaders; not exactly the *community of entrepreneurial employees* I have discovered in the Kairos Moment.

History records most idealism as short-lived. As the Project Apollo gave way to Skylab and Space Shuttle, Philco became Philco-Ford, then Aeronutronic Ford, and then Ford Aerospace. It subsequently sold off to where my stipend for retirement came from Lockheed Martin. Cost cutting and minority quotas became more important than idealistic quality goals. The public did not hold space flight as a high priority and cut funding. Quality standards were lowered in proportion to the budget reductions. These unenlightened decisions were direct contributors to the explosion that destroyed the Space Shuttle Challenger, killing its crew, but that's a subject for another day.

I left NASA in 1980 to bring the toolkit of knowledge that I had amassed from auditing hundreds of companies and by being a member of the Project Apollo team to the manufacturing community. I saw anecdotal and experiential information accumulate during my 14 years at the Johnson Space Center as valuable and practical for non-governmental organizations. Although I have yet to complete the effort by dissecting what I had learned and developed into a clinical methodology, I had the ability to analyze any given business process and offer an appropriate solution to productivity and quality problems. If I used the models I had developed for supplier evaluation at NASA in a manufacturing company, the organization would surely make dramatic increases in output, while significantly lowering scrap rates. My thesis was valid; right to the point of discovering that most companies did not have enlightened leaders with compelling visions and endless budgets to throw at problems. They did not have cadres of trained personnel on staff awaiting the challenge of fighting fires as they ignited. Finally, private industry could not afford double and triple redundancy in their systems, as we did at NASA to avoid 99.95% of any potential disasters.

The wisdom from the Kairos Moment is derived from capturing the key lessons learned (vision, mission, process, accountability) and applying them to the realities of each environment. That is why *how-to* cookbooks and quick fix case studies do not lead to systemic and lasting benefits. Only from enlightened leaders applying knowledge appropriately can wisdom and success be truly sustained.

CHAPTER 12
DO NO HARM

While the term "Above all, do no harm." is most often attributed to the Hippocratic Oath, it is actually just a behavioral precept taught to medical students. Do No Harm is a cultural tenet of the medical arts that, hopefully, roots in practitioners' behavioral standards where their diagnoses and care does not harm the patients they are charged with healing.

We have taken license and adopted the medical concept of Do No Harm (DNH) to manufacturing and service companies. In our experience, this has long been a missing tenet of business management. Unfortunately, there is no "oath" required for individuals manufacturing products or providing business services. While the DNH standard of care is reinforced with medical professionals over long periods of education and internships, entrepreneurs and business managers may or may not have any type of training in organizational standards of care. Even fewer have any training in risk mitigation. Among the hundreds of companies my associates and I have worked with for over 40 years, our universal conclusion is that the number of business leaders who objectively address any sort of social accountability for their stewardship in delivering products and services to consumers is tragically small.

For decades, we have worked with business leaders in providing proactive solutions for business process improvement, quality and reliability enhancement, and customer service excellence. Although we have implemented success cultures for visionaries, for many business people, the pain of change was often more acute than the pain of running a business the way we have always done. Today, many manufacturing and service organizations are experiencing the full fury of the recent business downturn. The proactive vision has been completely replaced with panic and self-protectionism. Each day, the press reports another round of layoffs. In our neighborhoods, restaurants close and businesses file for financial relief as frequently as they announced their openings and their profit numbers.

While manufacturing and service companies are suffering loss of market and revenue, the "business" of product liability continues to grow both in scope and impact. Lawsuits, product recalls, warranty claims, and consumers receiving defective goods are all on the rise. For a snapshot in time, thirty-five million consumer products were recalled in the United States in the summer of 2007. The Acting Chairman of the U.S. Consumer Product Safety Commission dubbed it the "summer of recalls." Civil litigation relating to those recalls continues to escalate. This was clearly symptomatic before the business bust of 2008.

This steady market growth in product liability comes from the lack of accountability of manufacturing and service companies and from consumers becoming more aware of potential remedies available to them for perceived and real harm. We have more than enough case studies in our files[27] to assure you that product and service liability is systemic and may even be epidemic, both from the perspective of the consumers and the manufacturers.

On one end of the spectrum, we have been told that a way to make some fast money is to stage a trip-and-fall accident in a large retail store and their insurance carrier will quickly pay you some amount of money as a "nuisance" to avoid potential litigation. On the other end of the spectrum, we have discovered business operators who build "an acceptable kill rate" into their balance sheet who knowingly sell potentially dangerous products. These unscrupulous individuals assume that very few consumers will complain, fewer will return defective products, and an even smaller number will actually be hurt by their products. I have witnessed companies that keep product liability attorneys on retainer to handle their predictable number of annual lawsuits.

Fortunately, the fraudulent accident victim will most likely not get their illicit windfall as insurance companies have become less prone to give away "nuisance" money. Unfortunately, the unethical business people who have little social accountability are more likely to get away with selling substandard products because consumers do "throw away" goods of low dollar value rather than demand warranty or liability remedies[28]. To compound this unfortunate reality, many manufacturers that sell through fulfillment centers or large retail chains never see the defective products. It is not uncommon to have some number of warranty replacement items included in each shipment because the cost of dealing with individually

---

27   See Chapter 4 – Headlines of Shame

28   I recently purchased LED light bulbs to replace the incandescent bulbs in a bathroom. The package claimed that they are warranted until 2024. I put a note inside the bathroom cabinet to remind me to outlive the warranty and test it to see if they will replace my bulbs a decade from now.

failed products is too expensive and the retailer is instructed to dispose of the defective items. We observed one manufacturer that had a "no questions asked" return policy to instruct the retailer to destroy the returned products so that they did not wind up in garage sales and diminished new product sales.

Most reckless manufacturers receive point-of-sale data about returned items, but dismiss the returns as buyer's remorse the product of irresponsible and incompetent consumers. Since the manufacturers do not perform failure analyses on returns they do not see, they are often oblivious to inherent defects in their products. This group of manufacturers and service providers will keep me busy as an expert witness for many years to come. This book is, however, targeted at business leaders who first have a social conscience, second, want to minimize their risks, third, are concerned about customer satisfaction, and fourth, do not want to see me as an adversary on the other side of a court room.

You may already have ticking time bombs of liability in the field waiting to explode in your face that will manifest at the worst possible time as a lawsuit. At the very least, in times of economic downturn, many people are looking to blame others for their financial woes and may assault your company with frivolous liability accusations in hope to finding a source of revenue in collecting settlements from your insurance carrier.

What we are recommending is not a new "program" or a book you must rush out to buy. Like creating a vision, Do No Harm is a cathartic process of examination and assessment of the underpinnings of your organization. It is a full-body MRI for your company. It is a compelling wake-up-call for you to do something about that chronic pain in your chest. Beyond the metaphors, it is a process for locating the potential sources of liability that are inbred or have unconsciously evolved in your design, processes, products, and services. Do No Harm is an appeal for you to look within your business culture, from a very non-traditional perspective, to analyze your processes and their interactions with parasites or cancers that have been unintentionally introduced as your organization has evolved.

| Coffee Mug I used nearly every day from 1969 to 1979 while working at Mission Control, Houston | Coffee Mug I use nearly every day since 1999 while working as a Senior Consultant at Dell Computer | Coffee Mug I use nearly every day purchased At Space Center Houston in 2011. It used to display the proud NASA logo and "Failure is Not an Option." After 18 months, even the Made in China logo has washed away. |

How Do You Define Quality?

THE FMEA AND FBP APPROACH TO DO NO HARM

Many business professionals are familiar with the process of Failure Mode Effect Analysis (FMEA). This is a process by which products are analyzed for potential failure modes where prioritized list is generated to work on driving potential failures from the products. The method we use is called Forensic Business Pathology© (FBP). It compels you to move beyond traditional FMEA methods to not only look at all apparent opportunities for failure, but to look at how customers may misuse, abuse, and use your products for purposes other than what they were intended to do. Instead of applying rules of logic, we look for symptoms of illogic. We perform timeline analyses of the evolution of products and services to find spurious events that may affect quality, reliability, or safety that entered the process and have gone undetected. FBP tears down processes into their most elemental states and looks at them in three dimensions with unfiltered glasses. Instead of taking any parameter for granted, we use a form of Boolean logic (and, or, else) to determine if process elements have any potential for misuse or misapplication. FBP then, constructs scenarios of the sequence and interaction of these processes and how unexpected outcomes can result from processes previously assumed to be stable and reliable. Most often, FBP uncovers irrefutable evidence of processes gone awry that were not detected with usual business metrics and analysis. Unfortunately, these investigations often uncover a breakdown in accountability and stewardship by one or more elements (people, departments, philosophy, unconsciousness, ambivalence) within an organization. The serendipitous alignment of one or more of these undetected problems, plus some unforeseen scenario from a customer, is often the catalyst for catastrophic failure leading to your organization being blind-sided with a lawsuit.

After FBP analyses are run through a complete cycle, the fundamentals for a new culture of Do No Harm can be codified and your organization can determine how to best drive potential liability systemically from your products and services. Again, DNH is not the program of the month, but a fundamental and immutable set of standards of practice and care that must become absolutely innate to conducting a business. This prescriptive language sounds like a doctor lecturing a patient who is recovering from their first heart attack, but it is just as compelling. The alternative is to take risks with your business until it collapses from warranty costs and liability suits. Yes. There are those companies that appear to continually avoid consequences while using questionable business practices. With the proactive tools available to you, however, are you going to live life on the edge of risk or develop a philosophy of Do No Harm and realize success is an inescapable result of social and organizational accountability? If you chose the former, I suggest that you go to your local casino, find the Roulette table and put all of your money on "red." That bet makes as little sense as ignoring foreseeable risk in your business.

Accountability Model
© 2009 The Taormina Group, Inc.

VOLUME VI
PRELUDE TO THE APOLLO BUSINESS MODEL (ABM)

Alas, the bad news is that it is too late to save the Holy Roman Empire. The good news is that the current-day Romans are doing quite well in spite of the catastrophic mistakes of their ancestors. The better news is that it is not too late to save The United States of America from self-destruction if we apply the wisdom gained by events such as the fall of the Roman Empire modulated by the awareness of the examples I have presented in this book of our modern version of historical (and inevitable) social decay. Although it may appear that we are repeating many of the mistakes that caused the fall of Rome, we can still correct them before more irreparable damage is done.

The morals and values that kept the Roman legions and the civilization together could not be maintained towards the end of the empire. Crimes of violence made the streets of the larger cities unsafe. Emperors like Nero and Caligula became infamous for wasting money on lavish parties where guests ate and drank until they became ill. The most popular amusement was watching the gladiatorial combats in the Coliseum, which were attended by the poor, the rich, and frequently the emperor himself. As gladiators fought, vicious cries and curses were heard from the audience. One contest after another was staged in the course of a single day. Should the ground become too soaked with blood, it was covered over with a fresh layer of sand and the performance went on.

During the later years of the empire, farming was done on large estates that were owned by wealthy men who used slave labor. A farmer who had to pay workmen could not produce goods as cheaply. Many farmers could not compete with these low prices; thus, they lost or sold their farms. This not only undermined the citizen, but also filled the cities with unemployed people.

After the reign of Marcus Aurelius, inflation became rampant. Once the Romans stopped conquering new lands, the flow of gold into the Roman economy decreased, although it was still being spent to pay for luxury items. This dried-up the supply of gold by using it to mint coins. As the amount of gold coins decreased, they became less valuable. To make up for this loss, merchants raised the prices on the goods they sold. Many people stopped using coins and began a barter system. Eventually, salaries had to be paid in food and clothing and taxes were collected on fruits and vegetables.

Wealthy Romans lived in a domus, or house, with marble walls, floors with intricate colored tiles, and windows made of small panes of glass. Most Romans, however, were not able to afford a domus. The poorer working classes lived in small "slum" rooms in "high rise" apartment houses called "islands." Each island covered an entire block. First-floor apartments were not occupied by the poor since the rent for these living quarters were too high. The higher the shaky wooden stairs a family had to climb, the cheaper the rent. These upper-floor apartments were similar to today's tenements. Anyone who could not pay the rent was forced to move out and live on the crime-infested streets.

It was my initial intent to draw parallels with these examples of ancient Rome and our current society in America. What's the point? They jump off the page at you. The only major difference I see is that the more expensive apartments, penthouses, are not on the first floor. That anomaly is probably due to more robust building construction.

It _is_ my intention to use the rest of our time together to present some specific actions we can take to prevent our empire from falling. WAKE UP AMERCIA!

Since the Kairos Moments, past catastrophes are off most of our radar screens and classically defined Kairos Moments do not happen every day. I have developed a tool called Kairos Learning Moments, or KLM's that teaches us to look at every problem as if it were a Kairos Moment, from which knowledge and wisdom can be extracted. To use KLM's effectively, we must first be aware that the drama of catastrophes forces us to pause in our daily lives to look objectively (and emotionally) at the events that have temporarily interrupted our routine. In solving everyday problems, we usually do not pause long enough to study the root causes that brought the problem about. I'm not advocating us to focus on the dynamics of the coefficient of friction while you are skidding on an icy road. I am suggesting that you pause, on the side of the road, after all safety issues have been handled, and quickly reconstruct the scenario that made you skid in the first place. Was the skid a result of inappropriate speed, worn tires, poor choice of road, and inattention to the weather forecast or were you distracted by your cell phone? The root cause(s) can be determined by just a few moments of introspective thought, analyzing the data presented to you, and taking positive actions to fix the problem(s). If you don't correct the root cause, the next time you skid, you may have plenty of time to reflect upon the "why" in a hospital bed.

If you are going to give this tool a try, you must also understand the difference between <u>causes</u> and <u>symptoms</u>. Skidding is a <u>symptom</u> of some combination of actions that failed to occur correctly. Worn tires, excessive speed, and talking on the cell phone may be the <u>cause</u> of your car breaking traction and you losing control. All three causes can be fixed with some direct corrective action, if we accept the accountability to fix the problem! Unfortunately, we most often react by contemplating how to better control a skid, rather than how to avoiding another one.

Our lack of discipline for determining root causes (or our laziness and apathy) is why we spend billions of dollars in the "cold medicine" aisles of the drug store instead of taking the time to analyze how our personal habits and hygiene can be modified to prevent the scenarios that allow us to contact colds and influenza. I submit to you that if we developed the Kairos Learning Moment tool for every significant problem in our business and personal lives, and acted quickly on the knowledge and wisdom we derived from them, we would eradicate apathy, mediocrity, welfare, and sub-optimal performance within a generation.

I learned to drive in New York City, which means that I am, by programming, continually in a state of ***situational awareness*** of every activity around me so that I am not broadsided. Especially when driving, I have an innate sense of my environment that is constantly on alert. This gift allows me to always look for problems, instead of symptoms. I am also a trained observer. My gifts and training notwithstanding, I still take cold medicine and occasionally skid on our icy driveway. It is a shame that my humanity still causes me to miss so many Kairos Learning Moments. We are all fully human, but learning this tool can be a great first step in fixing many critical problems so they don't continually recur. Let's step through some of the learning opportunities we've identified in the previous chapters, review some of the tools we have already discussed, and apply the Kairos Learning Moment tool. Perhaps we can take some first steps in beginning our implementation phase of It WAS Rocket Science with some powerful tool sets.

CHAPTER 13
ROOT CAUSE ANALYSIS

The following are examples of problem solving logic, root cause analysis and Kairos Learning Moments we have been teaching and promulgating in our work with enlightened business leaders.

Problem 1: An appliance repair company had to make two trips to fix a simple problem in our clothes dryer. We were inconvenienced by using a Laundromat for a week and believed that we were overcharged for inferior service.

Symptoms: The service technician was inadequately trained to diagnose the likely causes of problems with this dryer. He was not trained in customer service skills. He was less-than enthusiastic about performing his job.

Root Causes: The Company has inadequate standards of training and behaviors for their service technicians. The principals are probably unskilled in creating a work environment that attracts motivated service personnel. They may not have done the analysis on how much money they lose on repeat service calls and unhappy customers.

Kairos Learning Moment: Truly successful service businesses have learned to build a work force of enthusiastic and personable individuals who have above-average diagnostic skills and communications with the public. These enlightened organizations are more selective in their hiring practices because their technicians are representing the Company and its reputation every time they are on a call. They pay these individuals more than typical wages and train them in technical and human skills on a regular basis. They recoup those costs many times by saving diagnostic time, minimizing warranty calls, and creating happy customers who refer their services to others[29].

Problem 2: Our public school systems are graduating more functionally illiterate and ill-prepared students than ever our history.

Symptoms: A growing percentage of high-school graduates can't spell well enough to write basic business communications, can't perform basic mathematical calculations needed in daily life, don't know how to verbalize their thoughts in a form that is grammatically correct, are unable to interact with any degree of skill on current events in government and society that may affect their daily lives, have very little respect for any authority figure, have underdeveloped social skills interacting with others outside their age group, and have little sense of value or consequences on how their actions might affect those around them.

Root Causes: Educators have had to assume the roles of absentee parents. Students are sent to school lacking the discipline, respect, and social skills necessary to assimilate what they must learn to function in the world after they graduate from school. Parents are ill-prepared to deal with the outside influences that mold their children's value systems. Parents model lower standards of social accountability than ever before. Educators are overwhelmed with external influences (TV, computers, movies, music) that skew their student's values and negate the positive social and behavioral lessons usually learned in school.

Kairos Learning Moment: Schools perform to the levels of excellence demanded by the parents of the community. Whether by design or by abdication, schools have become a battleground between delivering curricula imposed by the school board and maintained enough decorum to have the information be absorbed by the students. The problems are seldom remedied systemically because the parents and educators focus on curing the symptoms and not permanently solving the root causes.

Problem 3: Users of Windows-based software spend an unacceptable

---

[29] We recently adopted a rescue Chihuahua. She is 7 lbs and just about a snack for our resident coyotes and mountain lions. For the very small project of building a dog run. One bid was $3,000, the second was $2,800 and proposed chain link fence and concrete footings. We've never gotten a follow up call from either of them. The third came from a contractor who has been the victim of my consulting and training. He did the job with poultry wire for $500 and then called three days later to make sure we were satisfied with the workmanship.

amount of time dealing with program crashes, error messages, lost data, user-unfriendly program quirks, compatibility issues, and program inefficiencies. Users also are exploited by being lured to continually pay for upgrades that they may or may not need.

Symptoms: Programs that do not install properly. Programs that crash for no apparent reason. Error messages that require extensive research to fix. "Users" have to become educated on software and hardware just to know how to diagnose recurring problems. Data loss and corruption for no apparent reasons. Programs that are designed for the common folks which are less than intuitive in their use and efficiency. The opportunity for reasonably skilled users to cause "fatal" errors which require extensive repairs. Waiting for the "hour glass" (slow processing speeds). Inadequate documentation. Help and repair tools that are ineffective. Inevitably, dealing with the "blue screen of death"[30]

Root Causes: With our insatiable appetite for "more and faster," we have created a culture within the hardware and software industries that has transferred responsibility for delivering robust and reliable products from the manufacturers to the users. Our lost time submitting problem reports provides the data to stabilize products that have been delivered, knowing that operational revisions are needed. Because there are limitless numbers of software providers selling Windows-based software and no established standards of compatibility that are enforced, conflicts and crashes are inevitable. Also, hardware manufacturers are not held to compatibility standards, which allows each PC to be an opportunity for unique hardware/software conflicts to occur, based on the endless permutations that software and hardware users may configure.

Kairos Learning Moment: By our insatiable demand for more features that perform more quickly with more capacity and higher resolution graphics, we have allowed ourselves to settle for products that are not completely developed by the manufacturers. In the case of Microsoft, our demand has created the richest corporation in the world that delivers products that are constantly in a state of problem correction. By accepting mediocrity as a norm, we have signaled the rest of the business world that American consumers will accept products that haven't been fully developed or debugged. This is manifest in the monthly recall of automobiles, food products, and consumer goods. It has expanded the need for

---

30   When a Windows computer experiences a fatal crash of its operating system and cannot be restored by an untrained individual.

consumer watchdog groups and the role of federal regulators. Is this the business model we want to continually expand and degrade? Do we need more fodder for product-liability law suits?

A couple of hours after writing the above scenario (Problem #3), my anti-virus software crashed and I spent the better part of three hours downloading updates, reading FAQ's, plodding through unhelpful-help menus, and uninstalling and reinstalling the software. Coincidence?

These three examples are offered to suggest how we can capture KLM's, format them in a style that separates symptoms from causes, and gives us an objective set of criterion from which we can develop an action plan. Now comes the hard part. Once we have identified the problem - the root causes and our expectations - how do we implement real and lasting changes that will not only solve a problem, but will create an environment where the same problem does not have to be solved repeatedly?

In the Introduction, I briefly acquainted you with our "4P" model (People + Process + Power = Performance) for building a high-performance business. I made it through the entire rest of the book without using another consultant-acronym, until now. There's just something about a former NASA guy that draws me almost uncontrollably to creating clever names and acronyms for the tools we use. After hours of work experience and brainstorming with my inner circle required to validate these methods, a cool name for our latest program is like a reward we must give ourselves. Heck, I just realized we've never named the form I used above to document KLM's. Maybe we'll have a naming contest....

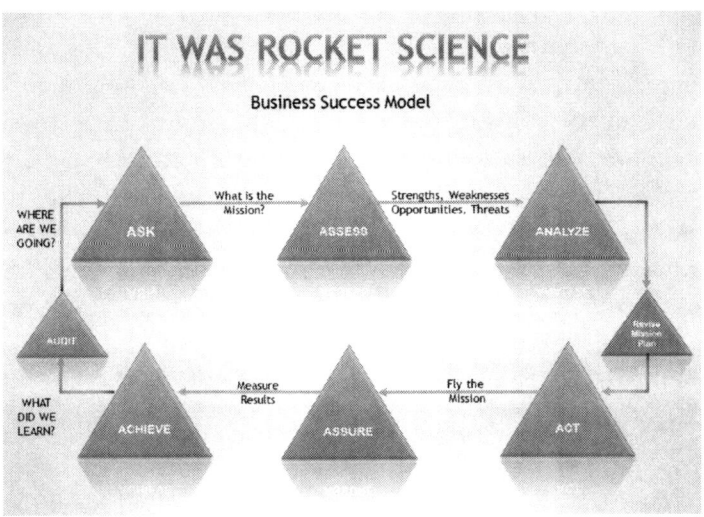

CHAPTER 14
SEVEN ALPHA

The tool above, which we have validated over a decade, is called Seven Alpha. It is the foundation of the Apollo Business Model. It is a hybrid of classic methods like Plan-Do-Check-Act, however, we've added a few steps that are critical to making it a self-help tool that you can use without the need for paying consultants. Business folks who have taken it to heart use this method to solve problems, from the mundane to the catastrophic, and never leave home without it. The reason for its success is that it forces the user to resist moving directly from symptom to resolution, as we are given to do. It keeps us from wasting time and resources on solutions that may be incomplete or causes the problem to recur. I recommend you try this Seven Alpha map the next time you have a business problem smacked you in the face.

1. ASK – The individual or group that identifies and proposes a problem resolution must have permission to expend the time and resources to perform the investigation, the power, and the accountability to implement the proposed solution. The approving authority must have committed to following the recommendations of the problem solver(s). They must agree to whatever milestones and approvals the authority has established. Risks and rewards must be agreed upon in advance. Grass-roots efforts can backfire without the authority to spend resources and ill-advised if the proposed solutions are rejected because the authority was not plugged in ahead of time. Many great ideas die from lack of planning and authority.

2. ASSESS – Determine how the stated issues can exist within the context of the organizational environment. Let me repeat that. An ab-

solute key to problem resolution is to determine how the problem has been allowed to exist within the processes, procedures, checks/balances, and oversight that currently exist. Problems are only problems because some combination of conditions has opened a chink in the armor that has allowed the problem to incubate and grow. What is the actual problem? Use the KLM format to separate problems from symptoms. Many times, the root-cause problem is totally hidden within the chaff of the symptomatic morass. It's like the earlier example about the common cold. Our body has to have some set of physical conditions out of order that allows us to be vulnerable for the cold virus to take hold. We have to identify these conditions and correct them before we can prescribe a long-term remedy. Another interesting component to the "virus" metaphor is that, once we have survived the disease, our bodies' build-up immunity to the virus. In time when our resistance wears down, we forget how difficult the illness was, yet, we allow that environment to exist and expose us to the same virus. An ounce of prevention is worth a pound of cure?

3. ANALYZE – Formulate solutions based on the objective analysis of the data from the first three steps. Build an implementation plan, schedule, and resource-needs assessment. Brainstorm the proposed solution with all who will be affected. Reach consensus that the plan is the best that you can propose, given the constraints of time, budget, and the potential rewards. Gain commitment from everyone to work the plan until the goals are achieved or measurable milestones must be developed because results did not match expectations.

4. ACT – Implement the solutions methodically, systematically, enthusiasm, and with velocity.

5. ASSURE – Measure key milestones against predicted results. Modify the plan based on metrics, not emotions. Have measures in place to know when you are "done" so the solution does not take on a life of its own.

6. ACHIEVE – Recognize success by presenting "before and after" analyses. Publicize the success and implement any deserved-rewards in a timely manner. Set follow-up processes in place to ensure long-term success. If the problem recurs, go back to Step #1. If not, **find a meaningful way to appropriately celebrate the achievement with all involved.**

7. AUDIT – Perform ongoing assessments of process effectiveness. What you observe in product and service quality does not necessarily reflect the effectiveness and profitability of you processes.

Let's test-drive the Seven Alpha tool on Problem #1 that we developed earlier into a KLM. Let's assume we are the top managers at the appliance repair company and have to resolve the problem caused by Mrs. Taormina's complaint about the incompetent service, two-week delay, overcharges, mess on the floor, and oil on the driveway (note the "Mrs." – my Irish wife is highly skilled at resolving these types of complaints). This is a particularly good example because, before we even start on a solution, we have to acknowledge that the service technician in question is the son of the owner. When nepotism is one of the variables, there may not be a logical solution to the problem. So, the first step is to decide the wisdom of launching a problem resolution exercise at all. As I quoted earlier from the movie War Games, the best solution might be not to play. Assuming the employee in question is subject to the same rules as everyone else in the company, let's move forward.

1. ASK – From the KLM, we have already determined that the root-cause of the problem goes beyond the technician having an off-day. The complaint we received is a combination of failures on the part of our managers that allowed the scenario to unfold so badly. The first step, then, is to gain concurrence from the owner on the root causes and obtain his approval to spend the time and resources to perform the necessary analyses.

2. ASSESS – Step two requires some quiet time and patient participants to uncover what allowed the problem to exist in the first place. Once we have the authority, we need to interview the technician in question about what exactly his take on the complaint and the various components of the problem were. He has to know that he is not being setup, but that his cooperation is vital to improving overall company performance. We also need to interview everyone else involved in the processes that failed. A terrific customer-satisfaction tool is to ask the complainant to be interviewed with the goal of correcting your internal processes based on the customer's input and concern. I always welcomed such phone calls from service providers. It makes

me feel like my concerns are being heard. Have we become too busy to train our people? Do we just assume our people understand our expectations of customer service? Have we ignored procedures already in place to ensure service calls are completed as we've prescribed? Are there incentives for the technicians to do things right the first time? Are there any consequences if they do not perform to expectations? Are the technicians answerable to each other for their level of performance? Until we get viable answers to these questions, there is little purpose in attempting to solve any of the symptoms.

3. ANALYZE – Diagnose how the root causes were manifested in the symptoms. Brainstorm how each malfunction was caused and precisely what steps will be required to keep the problems from recurring. Map a plan, budget and schedule for the steps of corrective and preventive action. Secure concurrence with those affected and approval from the process owner.

4. ACT – Implement the agreed-upon plan systematically as drafted. Monitor measurement points that you have established to ensure the schedule is valid, the budget is on track, and the desired results are evolving.

5. ASSURE – Monitor the data points to assure the solutions are working effectively. Make any necessary changes to the plan as events dictate. Collect "before and after" measurements and validate success.

6. ACHIEVE – Present the results to the owner, staff, and work force. Enlist the support of everyone involved in making the solutions permanent. Schedule follow-up audits to ensure that no new sink-holes have opened that would undo the good work that was accomplished

7. AUDIT – Continually assess processes for efficacy and opportunities for improvement.

This methodology is so basic to my work with client companies that I have difficulty understanding why it isn't universally embraced. At the same time, I am aware of how long it takes us, humans, to break bad habits and to learn new tools, especially when they take more energy and discipline to implement than our "shoot-from-the-hip" technique. I have enough anecdotal and experiential information about its success that I can vouch for it as a business tool without hesitation. I often admonish my clients to not try this at home because you can't hire and fire family members at will. At the same time, it can be a valuable tool for solving systemic problems in our personal lives if, we have great personal discipline and resolve plus a supporting environment. I still have problems sticking with the program when it comes to eating habits and exercise!

The final frontier is implementing these problem-solving techniques to tackle our societal and political epidemics. There are two huge barriers to reaching the "authorities" in these arenas: power and ego. Only the most enlightened leaders are open to sharing their power with those around them, knowing that they will be more effective when they empower others (Empowerment is one of the most misused and misunderstood words in our vocabularies). Most "leaders" hoard their power and are threatened by collaborative problem-solving methods such as those I use and teach. In the political arena, power and ego are such massive forces of nature that systemic change is only accomplished in a Kairos Moment, such as the creation of The Department of Homeland Security after 9-11. Instead of the two-party political system creating some balance in our approach to governing, it has raised a bumper crop of egomaniacs whose sense of self has caused them to behave like petulant children instead of stewards of our collective future.

If I have been effective in making the readers of this book aware of the clear and present dangers signaled by the examples I cite of social and moral decay within our society, perhaps we can each affect one simple systemic resolution to a problem in our own companies, families, or communities and create examples of how to not follow the fate of the Holy Roman Empire. Perhaps we can only personally influence the value system of one child or bring about a single positive change in our company. The elephant is eaten one bite at a time. When I see a Habitat for Humanity project underway, I am reminded that there is at least one successful program that is a coalition between business, government, and social agencies that created a formula where individuals, who have been at the effect of some personal or natural disasters, can regain their dignity by working with their neighbors to build homes that solve the low-income housing problem, one house and one family at a time.

The secret of success is that there is no secret of success.
Elbert Hubbard

VOLUME VII
IMPLEMENTING THE APOLLO BUSINESS MODEL

At last, we have arrived at the launch pad and are armed with the history and context so that we are able to consider a historically viable model by which we can run our organizations. The steps covered below are elegant in their simplicity, but monumentally challenging to implement successfully. These tenets took 40 years to validate and document. We are, after all, fully human. And 50 years later, after JFK was assassinated, there is new evidence that he may have been accidentally shot in the head by the unintentional discharge of a rifle, which belonged to a secret service agent that was behind his vehicle and sworn to protect his life.

Up until now, I have provided anecdotal information and a number of tools to add to your arsenal of enlightened leadership. Now comes the opportunity for you to systematically perform a catharsis within your pathological organization.

There are nine tenets to the Apollo Business Model. None of them is optional. None is more or less important than another. They must be assessed and implemented in the three tiers shown in the figure below, from left to right and from top to bottom in the sequence that follows.

Unlike the other tools presented in the book, this is a strategic business model. It is a culture, not a methodology or a management fad.

It is not unusual; when I present the model to business leaders, I am sent away when they realize that the fundamental values of it will not fit in their organizations. Most often, this disconnect occurs in family businesses where there is a hierarchy that is unwilling or unable to change autocratic rules. In other companies, an informal organization has been allowed to form that runs the day to day operation by manipulating the

The Apollo Business Model
Wisdom from History

We Just Made Up Our Minds to Go to the Moon and Then Did It!

Vision	Leadership	Metrics
Mission	Process	Consistency
Values	Boundaries	Achievement

power structure against the workers. Informal leaders are often more difficult to dethrone than the CEO because they are typically in control of key processes and keep the secret recipe to themselves, so that they are perceived as invaluable.

As you read and internalize the Apollo Business Model, you get a sense that this can never work in your organization. You may be right. This is not the quick fix or magic pill for organizational success. If, on the other hand, you have that maverick spark and the determination of the early astronauts, this might be the turnaround you are looking for in becoming the brand leader or market dominator in your industry.

VISION

Foundational ABM Reference – JFK's 1970 speech committing us to going to the moon in the decade of the 1960's.

The definition of vision is in the exquisite metaphor that I used at the beginning of this book - JFK creating a visualization that we were going to send men to the moon and return them home safely before 1970. The Country was disappointed that the Soviets had, to date, beat us in every

Vision

ABM Attribute	US Business Practice Today	ABM Principle	ABM Benefits
Vision	• Maximize profit • Minimize cost • Marginalize risk • Sell the business	**Shoot for the Moon** Create your organizational vision and be the brand leader.	• Provide products and services of exceptional value. • Fill the customers' needs and expectations. • Do no harm.

milestone of space travel. Kennedy painted a picture of a nearly impossible challenge that each of us could envision. We could see beyond putting an American just in orbit, but actually beating the Russians to the Moon.

A vision is a statement that startles you into imagining some impossible outcome that you can identify with or take ownership of. It causes you to say to yourself, "Wow, we can do that." and that germ of an idea stays with you like a rash that itches, until you do something about it.

That takes us to the belief that it takes visionary leaders to have a viable vision that can compel others to achieve previously impossible outcomes. There are very few John F Kennedy's among us today, and we've already been to the moon, but there are other impossible objectives awaiting us. You do not have to be born into wealth and power to become a visionary. Leaders are manifest from necessity, disaster (remember the Kairos Moment), geography, timing, and events that are serendipitous. We cannot recreate the events that led Michael Dell to building computers in his dorm room, in Austin, Texas at the precise moment in time when PC's were emerging as the technology that changed our world forever. We also cannot mix that with his natural charismatic leadership or the fact that he was wise enough to surround himself with people who could complement his marketing genius. The brand leadership that he created was contextual, serendipitous, geographical, and a certain amount of luck and timing.

You cannot create those scenarios from reading this or any other book. You can, however, take time for introspective reflection and ask yourself some really dangerous questions.

- Am I a leader or a manager[31]?
- Do I operate by command and control or do I engage others to be creative?
- Do I encourage entrepreneurism in my people or do they suffer for their mistakes?
- Am I reasonably successful but unfulfilled in what I have accomplished?
- Am I not meeting my personal objectives and am I sincerely looking for a proven approach that will work for me, in my environment, in the moment, and with the constraints I have in my life and organization?

Each of these questions are a variation of "Am I a leader or a manager?" If you are a manager, you are most comfortable with being "in control," manipulating your environment, and distrustful of people. You will likely never be a visionary leader. If you are a pragmatist that never stops to evaluate where you are and searches for external answers as to why you or your company are sub optimized, you will not likely make it beyond the step of creating a shared vision that will be embraced by those who you influence.

If you want to grow, you must look inside yourself and identify your values and motives, and then document them in a vision you want everyone to share. The good news, if you are motivated, is that leadership and shared visions are learnable skills; not from reading books or going to motivational seminars. You need a facilitator who truly understands leadership and will give you brutally honest answers to their assessment of your leadership style, qualities, and effectiveness. They must also be able to assess if you are a "student." A student is on a never ending quest for knowledge and wisdom and knows that their journey is never over. Each time a student grows, he or she applies the lessons they have internalized and tests them for efficacy. If they do not have a return on investment, the quest for knowledge continues with the same zeal as in previous journeys.

So what exactly is a "vision?" It is the verbiage that, when spoken to a member of your team, a client, or a supplier, they immediately know

---

31 Leaders create an environment that allows others to grow. A manager manipulates people.

your value set and can identify with it, just like how millions of people identified with the vision of Americans landing on the moon.

In your business context, a vision statement might be:

- In the next decade, we will achieve brand leadership and become the benchmark of quality for our industry.
- We will create the technology to enable us to land on Mars within a decade.
- We will ensure that our students are prepared to become productive members of society when they graduate.

Now, this vision cannot be created by the marketing department. It must come from your soul and you must be prepared to be the standard bearer who lives the vision every day. Without that commitment, it is so much fluff. Publicizing it without living it will create more problems than you have now.

Second, you must be able to create a set of metrics that will continually let everyone know how you are advancing on your target and goals[32]. In the first example, you must make a commitment to find the professional societies or organizations that conduct reliable market research that can accurately assess your performance against your competitors and peers. Those data sets must be public information so that everyone in your organization has ownership. You must communicate to everyone how their work affects "our" vision. In the same example, you need metrics to demonstrate quality as an objective measure, not a feeling. In manufacturing industries, this can decrease the number of products that reaches a customer with the goal being zero. In service industries, the vision metrics may be formal customer feedback mechanisms.

The second vision statement is the one that I think will bring this Country out of malaise, apathy, and mediocrity. It will create new technologies that will solve our energy, environmental, and scientific needs BECAUSE we are going to Mars, not because we are building a more efficient solar panel or electric car. **We still do not have viable fusion energy because there is no compelling vision for human space travel.** We have computers, medical breakthroughs, and a host of published spinoffs from the space program that would have taken decades longer to realize if we had not focused on the mission of going to the moon.

---

32 The final Apollo missions were scrubbed because we achieved all of our planned goals so quickly!

The third example has to be more than a school board's answer to solving their budget woes. Each example must have a tactical plan for short term results, a strategic plan for long term outcomes, an implementation plan and schedule with real milestones, and a set of metrics to validate whether the vision is being attained.

If your vision is to meet and exceed customer's needs and you have yet to define the acceptable level of customer service and haven't defined the parameters of how you are going to exceed their expectations, you are fooling yourself and setting your people up for failure.

The first time I was exposed to "The Nordstrom Way"[33], where the leaders specifically defined that their employees were to take all steps to ensure that their customers received the best possible service and trained them in their limits of power to execute the vision, I finally understood the term vision. There are hosts of stories about Nordstrom sales people. One is about a sales person who found a customer's wallet and tracked him down until it was returned. Another is of a sales person who found plane tickets in a dressing room, hailed a taxi, and took the tickets to the customer at the airport. Yet another was about a customer who wanted a specific shirt. The store did not have the exact shirt so the sales person went down the street to a competitor, bought the correct shirt, and gave it to the customer because he was inconvenienced. Is this the vision you have communicated to everyone who works for you?

It is more likely that you have not codified your values and shared them with the people who represent you in the world. Each time one of your representatives takes an action to "meet or exceed customer expectations," they are setting themselves up for scrutiny as to whether they did enough, not enough, or too much to achieve customer satisfaction.

Are you aware of whether or not you have a vision? If so, have you taken the time to codify it, establish metrics, communicate it to all who you influence, measure it, and continually improve it? You can't move to the second step until this is done and it is REAL.

MISSION

Foundational ABM Reference – Each endeavor into space had a clear and concise mission plan that everyone understood.

I'll make a disclaimer at this point. If my use of terms, like mission and vision, are not exactly what you learned in B school or at manage-

---

[33] A book written by the founders of Nordstrom's high end department store

Mission

ABM Attribute	US Business Practice Today	ABM Principle	ABM Benefits
Mission	• Meet profit targets • Raise sales numbers • Penetrate more markets • Reduce overhead • Survive	**Liftoff to Splashdown** Charter the mission your company must undertake.	Provide products and services that are of such great value, reliability and safety that customers will demand them and profits become an inevitable outcome.

ment seminars, please understand that I have defined them in a context in which I live and teach my clients from my experiential database.

Mission is the path you have cast and have made known of the direction you will take in order to achieve your vision. My favorite example is a vision statement that took me and the senior staff of this company days to write. Yes. We spent several long days coming up with this vision:

Our company will manufacture and sell the finest microwave data radios in the world.

How can we have spent countless man-hours coming up with such a simple sentence? That seemingly fluffy statement went through dozens of iterations and had to meet my test of real goals, measurable outcomes, and language that any customer or employee could internalize. What is behind the sentence is the following:

- Our company – we will continue to grow our company and keep it out of the hands of corporate raiders.
- Will Manufacture – we will expand our manufacturing facilities to keep pace with our planned growth. We will not outsource any manufacturing that is not controlled within our corporate umbrella and quality processes.
- And Sell – we will maintain our own sales and marketing resources and not sell our products through third party distribution.

- Microwave – we will stay within our core competencies and not branch out to other radio spectrums.
- Data Radios – we will not build radios for voice or television applications - only data.
- In the World – we will expand our domestic presence first into Europe and then into a global market.

I use this example with every client I have helped in developing a mission statement. It is clear, concise, measurable, and not open for interpretation. It is the vehicle by which their vision statement is executed. It is a living document that is reviewed at every quarterly meeting. It is explained to all new employees along with the challenge that, if you live by the vision and mission, you will grow and prosper along with the company. If you chose any other path, we will help you find employment elsewhere.

Creating a mission for your company is the next step after you have codified your vision and values. This is not optional, not negotiable, and not a marketing ploy. Until you are prepared to make this commitment for yourself and your people, you will not find the most rewarding path to sustainable success.

VALUES

<u>Foundational ABM Reference</u> – We will not cause an astronaut to lose his life in space.

There may be an apparent overlap between vision and values, but this topic is too important to assume we've covered the subject. In your mission, you communicate what is important to you and your organization. Values are the metrics you hold dear. Money is the first to come to mind, but focusing on gaining revenue is not the means to becoming wealthy.

If we create the shared value that profit is the inevitable result of satisfied customers, then the creation of a defined and accepted standard of customer satisfaction, loyalty, and service becomes our value set. If we create a cadre of highly motivated, highly successful, and eager workers who are accountable for themselves and to each other, profit becomes the inevitable result of lower labor costs, elimination of rework, and elimination of customer defects. If our value system is to return community rewards for supporting your organization and helping you succeed and grow, then identifying and measuring those contributions and creating a

Values

ABM Attribute	US Business Practice Today	ABM Principle	ABM Benefits
Values	• The 30 day phone call with the Board • Acceptable defect rate • More money • Less work • Buyer beware	**Mission Objectives** Create value by producing products and delivering services of exceptional value	The values of the leadership team are defined, modeled and everyone is held to the same standards.

defined level of community support becomes another avenue to grow the business. If you endow the local university, your value system becomes one for creating your own future workforce of highly motivated individuals who have benefited from your donations.

Finally, values is what is shared with everyone in the organization. If punctuality is a value, then everyone must be held by the same standard. If minimizing lost time by encouraging healthy living is a value, it must be available to everyone. If entrepreneurism is a value, then everyone must learn risk, reward, and accountability.

If you get nothing else from this topic, personal accountability and the accountability of each person to the other has the greatest return on investment.

LEADERSHIP

<u>Foundational ABM Reference</u> – Those who led us to conquer the moon and space were vested in their vision, mission, and values and created an environment where we could achieve impossible tasks on time and within budget.

I have covered leadership and its tenets in other chapters. Once you have the vision, mission, and the processes to implement them at every level in your organization, you must become the model for both. You must walk the talk in everything you say and do. I had one client whose partners claimed to have mastered vision and mission and accepted their roles as leaders. Unfortunately, one of them had a detail shop come to the parking lot of the business and wash and clean his vehicle every week, while he told his people there was not enough money to air condition the assembly building. The other partner was buying parts for their products using his credit card for purchases and overtly using the reward points to play different golf courses around the country, again professing a tight budget that did not permit luxuries in the plant. They were not setting the example they wanted their people to live. The problem was not that they were reaping perks of their position, but they were flaunting it in front of their people who were holding tasks to streamline operations and reduce waste.

The good news is that leadership can be an acquired skill. As southern humorist Brother Dave Garner used to say, "You don't have to watch what you do. You have to watch what you think!"

Leadership

ABM Attribute	US Business Practice Today	ABM Principle	ABM Benefits
Leadership	<ul><li>Exploit people</li><li>Manage supply chains</li><li>Treat customers as a commodity</li><li>Do unto others before they do unto you</li><li>Abdicate accountability</li></ul>	**The Mission Control Team** Create an organization of leaders of people, and managers of processes.	Eliminate hierarchies allowing everyone to succeed and grow.

PROCESS

Foundational ABM Reference – Every action related to manned spaceflight was a well-documented process and the interrelationship of these processes was defined, tracked, and measured.

Every human activity is a process. It has a beginning, an activity, and an end. Then, another process begins. This concept is so fundamental to successful businesses, yet it appears that chaos management is often the preferred method of taking a customer's order to delivering a product or service. The following are examples of the concept of managed processes.

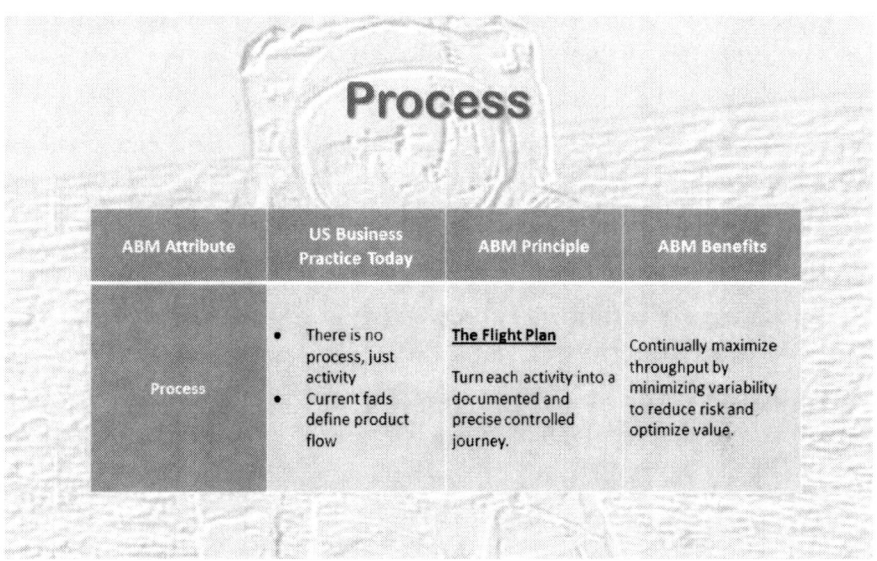

ABM Attribute	US Business Practice Today	ABM Principle	ABM Benefits
Process	• There is no process, just activity • Current fads define product flow	**The Flight Plan** Turn each activity into a documented and precise controlled journey.	Continually maximize throughput by minimizing variability to reduce risk and optimize value.

Process #1: Making an Uncle Gino's Pizza-to-Go Pizza:
A Typical Business Process:
1. Hire some high school students who will work for minimum wages.
2. Show them how to make a pizza a couple of times. Yell at them when they screw up.
3. Spend the rest of your career answering customers' complaints and interviewing new students to replace those who couldn't follow your simple rules.

A Well-Documented and Executed Business Process

1. Receive telephone order and type information into the data entry terminal. Confirm ingredients, size, and style with customer. Verify address and delivery location. Confirm price with customer.
2. Wash hands and put on a new pair of latex gloves.
3. Referring to the data entry, readout, select the size, and style of crust.
4. Brush two ounces of olive oil evenly across the crust.
5. Apply five, one-ounce ladles of tomato sauce evenly across crust.
6. Select toppings and amount from the data entry readout and apply evenly per the sample photos.
7. Place on the bake oven conveyor.
8. Print customer receipt and staple to a box for the size pizza ordered.
9. Stage the box at the end of the bake oven conveyor.
10. Fill the remaining order and check for accuracy
11. Locate the customer's address in the map book and note the location on the delivery ticket.
12. As soon as the conveyor delivers the pizza, cut into sixteen sections and box. Place the box in the hot pouch for delivery.
13. Check the order again for accuracy.
14. Drive to the customer's location. Obey all traffic laws.
15. Review the order with the customer to ensure accuracy. Verify the delivery time with the customer and note it in the travel log.
16. Count the money received and change given. Return to the restaurant.
17. Clean the hot pouch and return it to the delivery staging area.
18. Follow up in two hours with a phone call to ensure that the customer was satisfied with the pizza.

Most business would take the approach defined in "Typical Business Process." That is, most companies do not take the time to define processes so that the critical path of making a product or producing a service is clearly delineated. The business owner would typically assume that everyone understands what a pizza looks like and how one is put together. He or she would also assume that their employees would "just know" about personal hygiene, cash control, driving a delivery vehicle, and customer satisfaction. "After all, we've all had pizza delivered. Haven't we?"

In eighteen steps, we have taken the task of making and delivering a pizza from a mindless act of associated events with variable outcomes to an audit-able, repeatable manufacturing process that, when followed, ensures the customer receives the exact product ordered on time and the agreed upon price. By documenting a rote procedure, we have created a process that sets the stage for repeatability, measurable results, and critically identifiable process steps. If the process operators are trained in the procedure and held accountable for its outcome, then the little time it took to document the procedure can be paid for in consistency, productivity, and repeatability for customer business. When we know the steps we are taking, we can change them and improve them.

I challenge you, the next time you are inside a successful fast food franchise, pause and look at the workflow and processes behind the counter. Each step typically has a work instruction and visual representation of the required outcome permanently affixed to the work station. Process variability is not an option in your Jumbo Jack.

Obviously, documenting the procedure, by itself, will not motivate a minimum wage teenager to grasp your vision and deliver a perfect pie every time. We could enforce the procedure by standing over the employees and yell at them every time they miss a process step. We could dock their pay each time a customer complaint is received. We could, instead, use internal auditing to ensure our success. In the Gino's Pizza-to-Go example, co-workers would be trained to systematically audit the eighteen steps of the procedure. That is, they would examine the process, step by step, find evidence of compliance, document the results, and use the audit findings for corrective action and continual improvement, but more about the auditing process later.

It is so critical that we understand the concept of "processes" that I want to share one more example of how poorly most of us define and document it and how we are constantly disappointed in the outcome. Anyone who has children or was a child can relate to this example of the benefits of a well-defined process.

Think about how many times you've told your children to "Mow the grass." Were the results you obtained exactly what you expected? If you do not have children, think about the same scenario with your parent telling you to "Mow the grass." How many times were there missed spots, clippings not bagged, the mower left in the driveway, or the edging not done? Who trained you and your children to "Mow the grass?" Hasn't it mostly been by assumed observation and some intuitive sense of what a properly manicured lawn looks like? What would the process of "Mow the grass." look like if it were properly documented? This time I will include the vision.

Process #2: Mowing the Grass

Vision: To maintain a consistent appearance in the vegetation surrounding the house and to ensure a safe home environment, free from pest and rodent infestation.

1. Put on heavy boots and safety glasses.
2. Inspect the mower. Verify that the air filter is clean and the blade is tight.
3. Be sure the level adjustment is set properly and all wheels are of the same height.
4. Check the engine oil. Add if necessary.
5. Fill the gas tank. Go get gas if there is none in the gas can.
6. Start the mower and mow in a logical sequence.
7. Be sure to cut all areas that you can reach with the mower.
8. Inspect the edger. Add gasoline/oil mixture. Check to be sure there is enough cutting line in the edger and that it is properly wound.
9. Run the edger around all perimeter areas. Be careful not to damage flowers, trees, or decorative trim.
10. Rake up excessive grass clippings. Place clippings in compost bin.
11. Inspect your work and verify the job was done correctly. Report any damage that occurred during mowing and edging.
12. Clean the edger and mower and return them and all tools to their storage spaces. Report any observed problems with the mower or edger.
13. Keep track of when the oil needs changing in the mower or there is signs of blade wear.
14. Report any observed problems in the flowerbeds, the trees and shrubbery, or with insect infestations.

Do we communicate our processes or do we assume that people understand what is expected of them? If we are to be held responsible for our actions, they must be clearly communicated. They should not be complex in order to clearly communicate the expected outcome to a trained operator. If you can make your process procedures less complex or wordy than The Ten Commandments and as easily understandable, you've got an auditable process! Also, beware not to document "what you wish you did." In documenting processes, there is a tendency to take the process to new heights of complexity without checking with the folks who do the work. Committees can develop marvelous procedures, which appear to solve long-standing problems when their only task was to document what was actually happening. When an implementation committee is empanelled to be the "continual improvement" task force, they should start with what is actually happening, then involve the process owners in optimization. You can't effectively do both at once. By writing more into a process procedure than is actually happening, you will turn off the process owners because they will assume you are using the opportunity to add more complexity to their work without involving them. There is often a good reason for the process to be working as it is, no matter how convoluted it seems. Be sure to separate process documentation from process improvement.

BOUNDARIES

<u>Foundational ABM Reference</u> – Everyone within the space program understood the boundaries we must work to achieve our vision, mission, and values.

With a shared vision and mission, effective leadership, defined values, and documented processes, you need one more set of guidelines to create an organization that has the potential to achieve brand leadership and stellar performance. Boundaries are defined and agreed-upon limits of behavior in your organization. A metaphor would be a two-lane paved highway. The surface of the road is crowned and there are ditches on either side. If you do not continually steer your vehicle, the crown will eventually cause you to leave the pavement and wind up crashing in a ditch. Even if you ride the centerline, sooner or later you are going to have to follow your traffic lane or you will cause a wreck. When I work with companies, one of the first exercises I do is called always and nevers. These are a set of objective rules that are derived from vision and values.

Boundaries

ABM Attribute	US Business Practice Today	ABM Principle	ABM Benefits
Boundaries	There are no boundaries.	**Rocket Guidance** • Define the acceptable behavior of ALL. • Define unacceptable behavior and consequences. • Live consistently within the boundaries.	• Quality and safety are never compromised. • Organizations flourish when everyone operates within agreed-upon boundaries of what we always do and never do.

Here's an example.

"Always" and "Nevers"

The following are absolute and immutable requirements of employees and representatives of Consolidated Mfg. Co.

Representatives of Consolidated Mfg. Co. will always:
- Be accountable for their own actions
- Conduct themselves with decorum and veracity
- Treat everyone with dignity and respect
- Be champions of safety and reliability
- Shield customers from liability
- Be responsive to customer input and concerns

Representatives of Consolidated Mfg. Co. will never:
- Violate the intent of this document
- Impose their will on others
- Allow legal or illegal substances to impair their actions
- Violate any civil or criminal laws while representing Consolidated Mfg. Co.
- Misrepresent themselves or Consolidated Mfg. Co.

Just as creating the vision, mission and values are difficult. Creating the always and nevers is an exercise to see what really makes you tick. Whatever these statements of boundaries are for your company, if you have everyone sign the document, and then enforce it equally and fairly, your company will become free of drama virtually overnight.

Warning: These boundaries apply to EVERYONE EQUALLY. NO EXCEPTIONS.

METRICS

<u>Foundational ABM Reference</u> – Every process and activity has performance measures with clear and concise limits and tolerances.

If you can't measure it, it doesn't exist. I don't recall which of my mentors beat this into my head, but it has never let me down in business or in life. This topic is discussed endlessly in American Society for Quality meetings and chat groups because most business leaders understand it intellectually but still make decisions by "gut" or instinct and seldom look to data to validate a pending decision before making it.

Former Secretary of Defense, Robert McNamara, is quoted saying that we lost the Vietnam War because we did not measure the outcomes of our efforts, like enemy body count! There is also a concept called the McNamara Fallacy. It goes something like:

Metrics

ABM Attribute	US Business Practice Today	ABM Principle	ABM Benefits
Metrics	• Sales • Profits • Bonuses	**Instrumentation** • Provide instruments to measure all performance parameters. • Use the data to make changes. • Collect data sets to manage processes and the business.	Each process immunizes your organization from disaster by operating within the measurable parameters.

The first step is to measure whatever can be easily measured. The second step is to disregard those that can't be easily measured or to give it an arbitrary quantitative value. The third step is to presume that those that can't be measured easily, really isn't important. The fourth step is to say that what can't be easily measured, really doesn't exist.

That exercise in obtuse logic may sound absurd, but we often shoot first, and then aim. Reactionary decision making is as close to instinctive behavior as humans possess.

Every process within a business must have a set of metrics, key performance indicators, or whatever you call a graph that continually measures outcome against expectation. Every business decision must be based on the reality of these metrics combined with your entrepreneurial senses. Figures don't lie but liars figure. Avoid being data rich and information poor.

The best way to avoid disaster is to know when it is getting close by analyzing a performance chart that has upper and lower control limits. Do not let the process get close to either one before taking corrective action.

CONSISTENCY

Foundational ABM Reference – The reason we were able to rescue the Apollo 13 crew after the service module explosion was because we knew every parameter, every function, and every piece of equipment on the spacecraft.[34]

Most process problems, leading to unplanned outcomes, is the result of process variability. Maintaining predictable steps in your processes does not stifle creativity; it makes processes repeatable and measurable so that variations do not skew quality and reliability.

One of my clients builds custom windows and doors. Large portion of their clientele are people who own very old Victorian homes in the San Francisco bay area. These homes are decades old, have often been remodeled, and have survived many earthquakes. There is an art to measuring existing windows and doors, taking the data, and converting it into dimensions for the fabrication shop to use in creating new "custom" windows and doors. When they arrive for installation, they must be an

---

34   The astronauts fabricated a carbon dioxide filter from components available on the space craft so that they could survive the return trip home when neither the lunar module nor the command module had similar filtrations systems.

Consistency

ABM Attribute	US Business Practice Today	ABM Principle	ABM Benefits
Consistency	• There are too many variables to waste time documenting processes. • Writing job descriptions is too confining.	**Fly the Mission** • Execute the mission by the plan. • Modify the plan based on data points and metrics.	Achieve quality, reliability and profitability concurrently.

exact fit, even though there is often a shift in the opening when the old window is removed. It is not "okay" to leave a window open while they go back to rebuild the window.

We have conducted many brainstorming sessions with the fabricators and installers. We are writing work instructions for documenting the "art" of building custom windows and removing variability where unpredictability is a daily occurrence. In other words, surprises on site have been anticipated to the greatest extent practicable and the installers have been exposed to every conceivable scenario of process variability, thus they are prepared for any eventuality.

If your business is one of consistent and repeatable processes, there is no reason that you should tolerate process variability as an acceptable outcome.

The Carbon Dioxide air purifier fabricated by the Apollo 13 crew from components on board the spacecraft and duct tape. Do you have an inventory of all of the products and resources in your organization?

ACHIEVEMENT

<u>Foundational ABM Reference</u> – After every successful space flight, we had "splashdown parties" where we (all) celebrated mission success.

Performance bonuses for executives are how most companies celebrate achievement. Perhaps they throw a company Christmas party for the working folks or distribute some stipend amount of cash, but most businesses I work with do not know how to maximize celebrating success.

First, everyone who contributes has to share the risks and rewards. I met a custodian at Dell computer who was looking at his stock portfolio in the lunchroom. He told me that he knew that by keeping the facilities clean and safe, it directly contributed to his stock-sharing program. He knew exactly how many more years he had to work to convert his stock options to a retirement income. That is celebrating achievement! Giving employee's gift cards and ball-game tickets have a very short half-life of impact and effectiveness.

Also, basing rewards on making quotas or goals is meaningless if there is no risk for delivering defective products, losing money because of waste, or creating unhappy customers. I work with a marketing firm where no one gets a paycheck. Each person is paid when they have created the outcome they promised the client and met their promised objectives. What a wonderful model of risk, reward, and achievement that virtually none of you will find appealing!

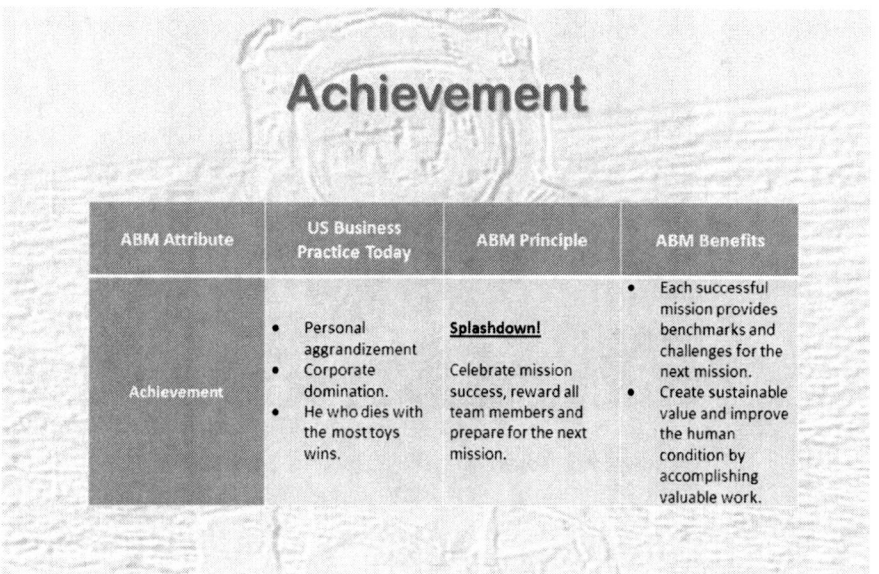

Achievement

ABM Attribute	US Business Practice Today	ABM Principle	ABM Benefits
Achievement	• Personal aggrandizement • Corporate domination. • He who dies with the most toys wins.	**Splashdown!** Celebrate mission success, reward all team members and prepare for the next mission.	• Each successful mission provides benchmarks and challenges for the next mission. • Create sustainable value and improve the human condition by accomplishing valuable work.

Splashdown Celebration for Apollo 11, the First Lunar Landing

CHAPTER 15
THE APOLLO BUSINESS MODEL SELF-ASSESSMENT

Do you have the right stuff? This simple assessment tool will help you objectively determine if you are a candidate to implement the Apollo Business Model.

ABM Attribute	ABM Principle	Your Score 0 or 1	Existing and Immutable ABM Value	Your Score 1 to 10
Vision	You have a published vision		Provide products and services of exceptional value. Fill the customers' needs and expectations.	
Mission	You have communicated your vision		Provide products and services that are of such great value, reliability, and safety that customers will demand them and profits become an inevitable outcome.	
Values	You have codified your values		Individuals who share the vision and mission of the leaders grow and prosper based on their individual and group accomplishments.	
Leadership	You lead people		Eliminate hierarchies allowing everyone to succeed and grow.	
Process	You have documented processes		Continually maximize throughput by minimizing variability to reduce risk and optimize value.	

Boundaries	You have defined acceptable behavior for ALL		Quality and safety are never compromised.	
Metrics	You collect data sets to manage processes and the business		Each process immunizes your organization from disaster by operating within the measurable parameters.	
Consistency	You operate to your procedures		Eliminating variability achieves quality, reliability, and profitability concurrently.	
Achievement	You reward all team members as you prepare for the next mission.		Create sustainable value and improve the human condition by accomplishing valuable work.	
		Principle Score		Value Score

ABM Principle

0 = You do not have the principle in place

1 = You have the principle in place and operational

ABM Value

1 = Something we would like to happen

10 = A foundational value that is not subject to compromise

_____ X _____ = _____

Principle Score x Value Score = ABM Self-Assessment Index

Self-Assessment Ranking

0 to 100   Your organization is fatally flawed

100 to 300   You have systemic problems that require immediate attention

400 to 600 Your organization has many opportunities for improvement

600 to 700 You lead your industry in many arenas

Over 800 You are the brand leader

© 2012 Copyrighted Property of the Taormina Group

CHAPTER 16
ONE COOLER TOOL

If I have managed to capture your attention this far, I will offer you a gift of one of our proprietary tools as a token of my appreciation. It is called the Organizational Liability Risk Index.

We use it in two ways. First, we furnish the tool to the senior managers at our client companies. We ask them to fill out the spread sheet and return it to us. Then, we conduct an on-site assessment of our own. The scoring system, itself, is very revealing of your organizational health. Comparing our score and your score is even more revealing of whether you are fully aware of the health of your company.

The tool is an Excel spreadsheet, so send an email to OLRI@itwasrocketscience.com and we will send you a copy. You are under no obligation to have us perform the onsite survey. We just ask that you do not duplicate or distribute the tool without our knowledge and permission.

Chapter 16 - One Cooler Tool

Organizational Liability Risk Index
A Numerical Rating of Business Accountability and Vitality

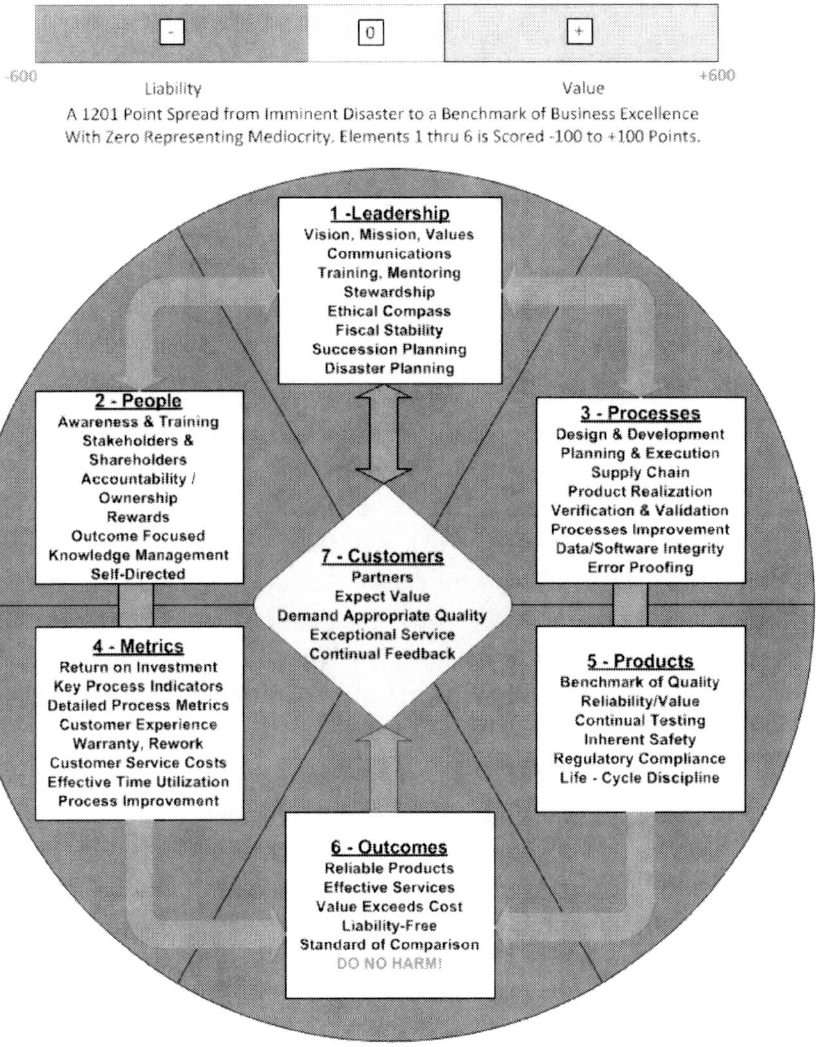

-600 Liability   Value   +600

A 1201 Point Spread from Imminent Disaster to a Benchmark of Business Excellence
With Zero Representing Mediocrity. Elements 1 thru 6 is Scored -100 to +100 Points.

1 - Leadership
Vision, Mission, Values
Communications
Training, Mentoring
Stewardship
Ethical Compass
Fiscal Stability
Succession Planning
Disaster Planning

2 - People
Awareness & Training
Stakeholders & Shareholders
Accountability / Ownership
Rewards
Outcome Focused
Knowledge Management
Self-Directed

3 - Processes
Design & Development
Planning & Execution
Supply Chain
Product Realization
Verification & Validation
Processes Improvement
Data/Software Integrity
Error Proofing

7 - Customers
Partners
Expect Value
Demand Appropriate Quality
Exceptional Service
Continual Feedback

4 - Metrics
Return on Investment
Key Process Indicators
Detailed Process Metrics
Customer Experience
Warranty, Rework
Customer Service Costs
Effective Time Utilization
Process Improvement

5 - Products
Benchmark of Quality
Reliability/Value
Continual Testing
Inherent Safety
Regulatory Compliance
Life - Cycle Discipline

6 - Outcomes
Reliable Products
Effective Services
Value Exceeds Cost
Liability-Free
Standard of Comparison
DO NO HARM!

THE TAORMINA GROUP

© 2010 The Taormina Group

EPILOGUE

Throughout the book, I have made a concentrated effort not to mire us in jargon or to fall into the trap of creating ideas for you that only work at Dell Computer or The Federal Reserve Bank. I've saved the following vignette for last to make the point that the tools presented herein work for mega corporations, for five-person companies, and, perhaps, for your community.

After avoiding the dreaded dentist chair for nearly five years, my anxiety level was quite high. I was expecting grim diagnoses of rampant diseases and monumental bills for amalgams and root canals. What I encountered was one of the most satisfying customer service experiences ever and an extraordinary model for running a small business operation. Instead of receiving and enduring an endless lecture on gum disease, I was witness to a living example of dynamic leadership that dismantles customer anxiety. I also experienced uncommon effective employee involvement and business process management.

When the subject of oral health arises, I try to understand what makes the strongest of us quiver at the thought of subjecting ourselves to the inclined chair of torture. Is it the submissive position, the bright light, or, perhaps, not being able to see the work being done? Could it be the sound of a high-speed dill in the next suite, laboring as it bores deeply into tooth and bone? Dr. John Aramini has developed an environment that can disarm the most anxious and reluctant patient. His dental practice has evolved into a pattern for total patient care, from the waiting room to the final reward of a new toothbrush.

As I reluctantly showed up for the appointment, the waiting room was full of other "victims", anxiously waiting their turn for torture. Behind the glass wall, I was prepared to find stoic, frocked munchkins who would prepare me for the prelude to pain with questionnaires and insurance forms. Instead, I discovered a spirited interchange of light humor among the front-office staff and other patients. At one point, the interchange became loud enough to override the soothing cadence of a mock waterfall in the waiting room. Realizing we were overhearing their conversation, Stephanie, the receptionist, poked her head out and asked, "You sure you want to come in here, we're all crazy?" I learned later that Stephanie has developed a knack for reading each patient and prescribing humor, empathy, and assertiveness, in the appropriate doses, to prepare them for the imminent trip to the dental chair. As I joined the front-office crew in dental humor and one-liners, my excursion to the inner sanctum was not as dreadful as I had anticipated.

Once I was poised for X-rays, Joy, the dental assistant, entered from stage left and continued the fun banter. As she described the upcoming events of my visit, I asked why everyone was so happy and upbeat. Her reply was wise in spite of her youth. She said, "We all love our jobs, our boss, and our patients." That answer was too general for me. I wanted to know more! I asked for an example. Joy replied, "We don't allow competition or drama in the office. We all leave our personal baggage at the door so we can have trust and genuine care about our patient's experience."

As the X-rays were developing, I was pondering Joy's words and preparing for the appearance of the obligatory hygienist to scale my teeth and perform gum torture with the polishing wheel. To my surprise, the dentist entered from stage right and introduced himself as "John," not Doctor Aramini. Instead of lecturing to me in dental babble, he began by creating a genuine relationship with me. While learning my concerns and interests, he would occasionally inject the diagnoses of the oral exam and the X-rays — in lay terminology and single-syllable words. After more explanations of the cleaning and soothing humor, he handed me the suction hose. I was to be in charge of my own destiny and insert the siphon whenever I felt the need. I would manipulate this instrument as an active participant in the process, instead of being a helpless victim. He then proceeded to clean my teeth with a new ultrasonic device that was totally non-invasive. Yes, the Dentist cleaned my teeth! If that wasn't shocking enough, his final comments were, "Well there's no decay or gum disease so, whatever you are doing, keep doing it!" How could he have departed so drastically from dental school lectures on brushing and flossing twelve times a day?

I was stunned that the appointment was over. As I was checking out at the front desk, I cornered John and asked if I could interview him and his staff to document a case study in customer service and process excellence. I had to know if John's success was merely a serendipitous accident or were he and Michael Dell cast from the same humanistic leadership mold. Obviously, he agreed to the interviews.

As it turns out, John had some very forward-thinking mentors at the University of The Pacific Dental School who did not subscribe to imposing professional arrogance onto reluctant patients. After starting his own practice, he began by making his own appointments, cleaning his own tools, and mopping his own floor. He also started building his bedside manner founded in the "chicken theory", a theory derived from the fact that he is personally uncomfortable sitting in a dentist chair. He explores every opportunity to make the experience as relaxed as possible for his patients, based on what disarms him when he is in the hot seat. With adults, he uses tact, compassion, humor, and his own admission that "some dental procedures just suck." With children, he encourages them to handle the dental tools and explains why the drill makes such an irritating whine. Actually, I found his explanation of how the drill works to be quite interesting.

During my interview with Brooke, the office manager, I discovered that she had spent the last nine years developing office processes based on common sense, group input, as well as some of the "chicken theory". She trains others by working with them through the language of dentistry and modeling customer service. Telephone and patient dialogue is a combination of diagnosing the needs of the patient, aligning with their state of mind, and providing services and appointments that cause minimal trauma. Since everyone has a hand in improving processes, the symbiotic effect has enabled the office to accommodate twice the number of patients that the average single-doctor practice can typically handle. In addition, patients generally do not return because they have moved, resulting in some 5,000+ patient charts occupying every inch of shelf space in Brooke's office.

What are the secrets of success of Dr. Aramini and his merry band? After my interviews, I now understand that Dr. Aramini's common sense success techniques for creating effective ways of dealing with people and processes is a "teachable model", applicable to any business.

First, John models and lives his values and vision. He is "hands-on" in training and setting the example for personal ethics, conduct, patient interaction, process execution, and corrective action. Everyone who works there quickly signs the customer service and internal work ethic model or they will "migrate out," as John puts it. While he has not wielded a mop in a while, he can still be seen helping with billing or filing and continually reinforcing his leadership-by-example.

Second, the doctor and his staff have created a unique work environment that requires everyone to "leave the drama at home." By living their internal business model for ten hours a day and not allowing internal competition or petty jealousies, there is minimal tension or miscommunication. Even when levity gets a little too heavy or a mistake is made, errors are treated as learning opportunities, and corrective actions are completed in minutes, not hours or days.

Finally, they jointly develop their processes and continually refine them to react to the needs of the moment. For example, in the last quarter of the year, they screened new patients to identify those who are in only for an annual insurance-paid checkup and those that are likely not to return. Every new patient is encouraged to become part of the practice, not a one-night-stand.

What lessons can be learned to help improve our businesses? They are elegantly simple:

- Share your vision and values with everyone
- Model excellence for customers and workers
- Treat employees as you want to be treated yourself
- Continually evolve business processes based on changing conditions
- Model personal accountability
- Leave the drama at home
- Continually put yourself in the shoes of your customers
- Wear a genuine smile, <u>especially when you have pearly white teeth</u>.

"Failure is NOT an Option"
— *Gene Kranz, Lead Flight Controller, Apollo 13*

Gene Kranz (Lower Right) Photo from the recovery phase of Apollo 13

CPSIA information can be obtained
at www.ICGtesting.com
Printed in the USA
FFOW04n1925220414
4947FF